Books are to be returned on or before
the last date below.

24 FEB 1990

R 2002

2

GUIDELINES FOR MONITORING INDOOR AIR QUALITY

GUIDELINES FOR MONITORING INDOOR AIR QUALITY

Niren L. Nagda, Ph.D.
Harry E. Rector, B.S.
Michael D. Koontz, M.S.

⦿ **HEMISPHERE PUBLISHING CORPORATION**
A subsidiary of Harper & Row, Publishers, Inc.

Washington   New York   London

DISTRIBUTION OUTSIDE NORTH AMERICA

SPRINGER–VERLAG

Berlin   Heidelberg   Paris   New York   Tokyo   London

1 2 3 4 5 6 7 8 9 0   B R B R   8 9 8 7 6

Library of Congress Cataloging-in-Publication Data

Nagda, Niren Laxmichand, date.
　　Guidelines for monitoring indoor air quality.

　　Includes bibliographies and index.
　　1. Air quality. 2. Air-Pollution, Indoor.
I. Rector, Harry E. II. Koontz, Michael D.
III. Title.
TD883.1.N34　　1986　　　628.5'3'028　　　86-9925
ISBN 0-89116-385-9　Hemisphere Publishing Corporation

DISTRIBUTION OUTSIDE NORTH AMERICA:
ISBN 3-540-17085-5　Springer-Verlag　Berlin

CONTENTS

PREFACE

Concern for air quality in homes, schools, offices, and other buildings has prompted research both on causes of indoor air quality problems and on potential solutions. In the last several years, our understanding of the multitude of contaminants present in indoor air and factors that contribute to poor air quality has improved considerably. Yet much work remains to be done to fully understand all pieces of the indoor air quality puzzle. The search for applications of this knowledge to improve indoor air quality is just beginning.

Advances along both fronts—causes and solutions—are predicated on measurements. Although the technical literature on indoor air quality is quite extensive on most matters, little direct guidance on measurement exists. The purpose of this book is to fill this gap by providing a systematic approach for monitoring indoor air quality.

This book is intended for individuals in government, the private sector, and academia. The government sector includes state and local agencies with responsibilities for public health, environmental quality, or resource conservation. In the private sector, the book will be useful to utility representatives, architects and builders, designers and engineers of building systems, industrial hygienists, and private consultants. University students, either at graduate or undergraduate levels, will find the book helpful in their pursuit of any of the types of disciplines stated or implied above. Researchers with experience in the field of indoor air quality may find the book a useful supplement to their existing information resources.

A number of individuals and organizations are to be acknowledged for their contributions at various stages in the development of this book. At the U.S. Environmental Protection Agency (EPA), Lance Wallace, David Berg, and Eugene Harris provided early support, especially for preparation of a report with the same title as this book, *Guidelines for Monitoring Indoor Air Quality* (EPA Publication No. EPA-600/4-83-046, NTIS Publication No. PB83-264465). This early work allowed us to organize design concepts in a form that can be useful to others.

Continuing support from the Electric Power Research Institute (EPRI) and the Gas Research Institute (GRI) has enabled us to broaden our thinking and our base of monitoring experience, leading to the refined and expanded background and guidelines that appear in this book. The authors are indebted to Irwin Billick of GRI and to Arvo Lannus, Robert Patterson, Ralph Perhac, Gary Purcell, and Cary Young of EPRI for these research opportunities.

Specific thanks are due to Katherine Hartnett (Brooklyn Union Gas), Jerome Harper (Tennessee Valley Authority), and Roy Fortmann (GEOMET), who, along with Irwin Billick and Cary Young, reviewed the manuscript and offered thoughtful comments.

Numerous manufacturers and developers of instruments provided detailed specifications on their instruments and shared with us valuable and sometimes unpublished information.

The authors are grateful for the support extended by GEOMET Technologies, Inc. In particular, we appreciate the invaluable word processing and editorial support provided by Laura Chen, Jean Fyock, Susan Perla, Jeanette Stalnaker, and Sharon Young of the Technical Support Department at GEOMET under the able leadership of Leonora Simon, our senior editor. Wayne Keyser, our photographic consultant, prepared for publication the photographs that appear in Appendix A.

Finally, thanks to our families who endured the many weekends and late nights that we spent preparing this book.

Germantown, Maryland   *Niren Nagda*
February 1986   *Harry Rector*
                                                               Michael Koontz

GUIDELINES FOR MONITORING INDOOR AIR QUALITY

Chapter 1
INTRODUCTION

It wasn't so long ago that indoor environments that did not involve heavy industrial activity were thought to be relatively free of air pollutants. At worst, the air quality in homes, schools, offices, and commercial buildings might have reflected outdoor conditions. Research conducted over the last two decades, however, has shown that pollutant concentrations in many of these indoor environments are frequently higher than outdoor levels suggest. Indeed, some air pollutants such as radon and formaldehyde are identified almost exclusively with the indoor environment.

Indoor air quality has become a significant environmental health issue because most people spend the great majority of their time indoors. As with outdoor air quality exposure and occupational exposure, monitoring pollutant concentrations is an essential part of evaluating potential health threats and identifying viable mitigation approaches.

However, measuring only pollutant concentrations trivializes both the question and the answer. That is, descriptive data that merely report pollutant concentrations are of limited value for decision-making unless underlying factors are considered.

Indoor air quality monitoring is usually initiated to address a specific type of research question or to respond to indications that air quality problems may exist in a certain building. The planning for such monitoring becomes complex as buildings come in many shapes and sizes with a variety of construction materials, furnishings, air handling systems, machines or appliances, and consumer products that can contribute to the problem. Further, most buildings are occupied or used by people with a multitude of characteristics and activities. These factors, in addition to indoor sources of contaminants, can ameliorate or exacerbate the potential problems.

Only a relatively small number of investigators who work routinely in this field are thoroughly familiar with the monitoring techniques peculiar to indoor air quality. The fairly rapid emergence of this area of research has led to the need for an instructional and training resource.

This book is written to provide the background, understanding, and guidance for developing a monitoring design that will adequately address a stated problem or objective. Chapter 2 gives a brief historical perspective on the topic of indoor air quality, highlights current areas of emphasis, and briefly describes different classes of pollutants that are common to indoor environments. Chapter 3 unveils the mass balance equation as a useful framework for understanding how various types of factors influence indoor air quality. Chapter 4 describes instruments and methods for monitoring indoor air quality and related parameters. Detailed descriptions in Appendixes A and B supplement the information presented in Chapter 4.

The heart of the book is Chapter 5, which describes a progressively intensive series of steps to develop a detailed design for indoor air quality monitoring. A number of examples are provided or referenced to illustrate the many considerations and principles that surround this task. Chapter 5 also outlines basic principles for investigating building-associated problems that have no clearly defined pollutant sources.

Chapter 6 supplements this discussion on design with practical considerations keyed to issues of probe placement and quality control. Chapter 7 lists some additional reading materials that can further supplement the knowledge base. Professional organizations that have been catalysts in the exchange of ideas and information on the subject are also identified in Chapter 7. The text and format of this book follows conventions of the International System of Units (Le Systéme International--SI). SI units and symbols that appear in the text are explained in Appendix C.

The complexities inherent in monitoring indoor air quality make it an exciting area of research where a great deal of the science and art of innovation can be applied.

Chapter 2
BACKGROUND

This chapter provides a brief overview of past and ongoing indoor air quality research and summary descriptions of pollutants commonly found in indoor environments. The overview and descriptions are not exhaustive, but provide some background information to readers who are unfamiliar with indoor air quality research or the sources and behavior of indoor pollutants.

OVERVIEW OF RESEARCH

The first modern studies on indoor air quality, conducted in Europe and the United States in the mid-1960's and early 1970's, measured indoor concentrations of pollutants that had been previously measured outdoors. Among the pollutants studied were total suspended particles (TSP), sulfur dioxide (SO_2), and carbon monoxide (CO) (Biersteker et al. 1965; Yocom et al. 1971). These early studies, as well as more recent efforts, demonstrated that indoor levels of an outdoor pollutant are affected both by outdoor levels and by indoor generation or removal. For example, indoor concentrations of CO are dependent on outdoor levels and on the extent of emissions from unvented combustion appliances within a building. On the other hand, in the absence of indoor sources, pollutants such as ozone (O_3) and nitrogen dioxide (NO_2) can rapidly decay indoors. Because of the importance of indoor generation and decay, indoor air quality research expanded to address indoor sources (Cote et al. 1974) and sinks (Spedding and Rowland 1970).

Although early indoor monitoring studies focused on pollutants governed by ambient air quality standards, the monitoring of contaminants that are primarily present indoors also began at about the same time. For example, an early study to quantify indoor levels of radon (Rn) was undertaken for the U.S. Atomic Energy Commission in the late 1960's and early 1970's (Lowder et al. 1971). Studies in Denmark in the early 1970's (Andersen et al. 1974) also identified formaldehyde (HCHO) as an indoor contaminant. In addition to the pollutants cited above, subsequent research projects have investigated contaminants such as inhalable particulate matter (IP), fibrous and biological aerosols, pesticides, and volatile organic vapor compounds (National Research Council 1981).

The infiltration of outside air through the building envelope influences indoor pollutant concentrations. Initially, studies of air infiltration focused on its relation to energy consumption; air infiltration is an important component of the heating and cooling loads of buildings. Beginning in the early 1970's, air infiltration was included as an important facet of indoor air quality monitoring (Drivas et al. 1972).

In early research, the ratio of indoor to outdoor concentrations of some pollutants was thought to be useful in predicting indoor concentrations (Benson et al. 1972). In the mid-1970's this ratio was replaced by a more fundamental mass balance approach (Shair and Heitner 1974). The mass balance modeling approach, simple in concept, was adapted from odor modeling in industrial hygiene (Turk 1963). In the mass balance approach, which will be described in Chapter 3, all factors that have an impact on the indoor concentration of a pollutant are considered in evaluating pollutant concentrations. In addition to improving predictive capabilities, the mass balance model permits a better understanding of the various factors influencing indoor air quality.

It was soon recognized that the total exposure of individuals to pollutants could not be adequately characterized by relatively sparse outdoor and indoor measurements. Instead, total exposure estimates had to account for all types of environments that people encountered in the course of their daily activities. Initial estimates of total exposure were synthesized (Fugas 1975; Moschandreas and Morse 1979) by combining "time budgets," or records of time spent by population subgroups in various locations, with concentration distributions that were measured independently in different types of locations or microenvironments.

The miniaturization of monitoring equipment, which permits pollutant measurements with devices that are readily portable or attachable to clothing, started in the mid-1970's (U.S. Environmental Protection Agency 1979a). In the last several years, the development of such devices has accelerated, and in some instances, the current state of the art in personal monitoring compares favorably to the technology of larger, stationary monitoring equipment. The advent of personal monitoring has encouraged research on total human exposure that includes measurements at home, at work, outdoors, while commuting, and in other environments.

Decreases in costs and improvements in sensitivity have made it attractive to use portable/personal monitors in a stationary sampling mode to characterize indoor environments such as residences. These smaller devices are sometimes used alone or in combination with more sophisticated equipment. The types of equipment that are currently available for measuring indoor pollutants and related factors are described in Chapter 4.

Thus, research in indoor air quality has steadily expanded in terms of the number of pollutants under investigation, the types of environments and factors studied, miniaturization of monitoring equipment, the level of measurement detail, and the mathematical analysis of results.

Indoor air quality research underway during the mid-1980's can be classified into five general areas:

● Research in chamber or laboratory environments (Cole and Zawacki 1985; Traynor et al. 1985) to characterize pollutant emission rates from a variety of indoor sources or to develop and evaluate pollution control systems or monitoring devices

● Monitoring in test houses (Nagda et al. 1985; Maki and Woods 1984) to study and simulate the indoor behavior of pollutants under different types of controlled conditions

● Field studies to characterize personal exposures of the general population (Akland et al. 1985) or specific population subgroups (Nagda and Koontz 1985), to characterize indoor environments (Leaderer et al. 1984), or for epidemiologic research (Ferris et al. 1979; Ware et al. 1984)

- Ad hoc monitoring in specific buildings in response to occupant complaints (Ferahian 1984; Morey et al. 1984)

- Development or application of theories concerning spatial and temporal variations in indoor pollutant concentrations (Sandberg 1984; Skaret 1984).

In the United States, indoor air quality research is sponsored by Government and private sectors. The Government sector includes Federal agencies such as the U.S. Environmental Protection Agency (EPA), the Department of Energy (DOE), and the Consumer Product Safety Commission (CPSC), State agencies such as the New York State Energy Research and Development Authority (NYSERDA) and the California Department of Health Services, and, to a lesser extent, local government agencies. Various gas and electric utilities and, especially, utility-funded research organizations such as the Electric Power Research Institute (EPRI) and the Gas Research Institute (GRI) have sponsored substantial research in indoor air quality, comparable in dollars to Federal-Government-sponsored research.

POLLUTANT DESCRIPTIONS

This section summarizes information on 13 of the most commonly found indoor pollutants and pollutant groups. Specific references are not cited, but references given at the end of this chapter as well as books and documents listed in Chapter 7 can provide further detail.

Sources and recommended exposure guidelines for these pollutants and pollutant groups are listed in Table 1. In the absence of a legislated or uniformly accepted guideline on pollutant exposures, the levels given in Table 1 have been compiled based on published information from different organizations or governmental agencies. The information listed in Table 1 is current to 1985; the bases for these guidelines are under review by cognizant organizations (Janssen 1984).

Asbestos and Other Fibrous Aerosols

Asbestos, which is a group of inorganic silicate mineral fibers, is a widely used component of school, residential, and private and public structures. The indoor release of asbestos depends on the cohesiveness of the asbestos-containing material and the intensity of the force that disturbs the asbestos-containing material. For example, friable asbestos, in the soft or loosely bound form used in fireproofing, can become airborne easily by a disturbance of the material surface. Hard asbestos-containing materials, such as vinyl floor products, release asbestos only upon sanding, grinding, or cutting. Studies have shown that indoor fiber counts and mass concentrations may exceed those outdoors, and on occasion the levels may approach the occupational standards (2 fibers/mL). During normal use, buildings containing asbestos have not shown higher fiber counts than are found outdoors. Limited data apply mostly to schools and a few office buildings, but the general public's exposure to asbestos fibers in public buildings appears to be low.

Biological Aerosols

Considerable evidence indicates that a number of organisms that cause contagious diseases--including those associated with influenza, Legionnaires' disease, tuberculosis, measles, mumps, and chicken pox--are capable of airborne transmission in the indoor environment. Respiratory problems such as common colds and pulmonary

TABLE 1. Sources and Exposure Guidelines of Indoor Air Contaminants

Pollutant and sources	Guidelines, average concentrations
Asbestos and other fibrous aerosols	
Friable asbestos: fireproofing, thermal and acoustic insulation, decoration. Hard asbestos: vinyl floor and cement products, automotive brake linings (0).*	0.2 fibers/mL for fibers longer than 5 µm (based on American Society of Heating, Refrigerating, and Air Conditioning Engineers (ASHRAE) guidelines of 1/10 of U.S. 8-h occupational standard (ASHRAE 1981)).
Biological aerosols	
Human and animal metabolic activity products, infectious agents, allergens, fungi, bacteria in humidifiers, bacteria in cooling devices.	None available.
Carbon monoxide	
Kerosene and gas space heaters, gas stoves, wood stoves, fireplaces, smoking, and motor vehicles (0).	10 mg/m^3 for 8 h and 40 mg/m^3 for 1 h (EPA 1971, 1975).
Formaldehyde	
Particleboard, paneling, plywood, carpets, ceiling tile, urea-formaldehyde foam insulation, other construction materials.	120 µg/m^3 (based on Dutch and West German guidelines as reported in ASHRAE (1981) and the National Research Council (NRC) (1981)).
Inhalable particulate matter	
Smoking, vacuuming, wood stoves, fireplaces, combustion sources (0), industrial sources (0), fugitive dust (0).	55 to 110 µg/m^3 annual, 150 to 350 µg/m^3 for 24 h (EPA 1981).
Metals and other inorganic particulate constituents	
Lead: automobile exhaust (0).	1.5 µg/m^3 for 3 mo (EPA 1978).
Mercury: old fungicides, fossil fuel combustion (0).	2 µg/m^3 for 24 h (ASHRAE 1981).
Cadmium: smoking, use of various fungicides (0).	2 µg/m^3 for 24 h (ASHRAE 1981).
Arsenic: smoking, pesticides, rodent poisons.	None available.
Nitrates: outdoor air.	None available.
Sulfates: outdoor air.	4 µg/m^3 annual, 12 µg/m^3 for 24 h (ASHRAE 1981).

(Continued)

* (0) refers to outdoor sources.

TABLE 1. Sources and Exposure Guidelines of Indoor Air Contaminants (Concluded)

Pollutant and sources	Guidelines, average concentrations
Nitrogen dioxide Kerosene and gas space heaters, gas stoves, combustion sources (0), vehicular exhaust (0).	$100\ \mu g/m^3$ annual (EPA 1971).
Ozone Photocopying machines, electrostatic air cleaners, outdoor air.	$235\ \mu g/m^3/h$ once a year (EPA 1971, 1979b).
Pesticides and other semivolatile organics Sprays and strips, outdoor applications (0).	$5\ \mu g/m^3$ for chlordane (NRC 1982).
Polyaromatic hydrocarbons Woodburning, smoking, cooking, coal combustion (0), and coke ovens (0).	None available.
Radon and radon progeny Diffusion through basement floors and walls from soil in contact with a building, construction materials containing radium, untreated ground water containing dissolved radon, radon from local soil emanation (0).	0.01 working level annual average (ASHRAE 1981).
Sulfur dioxide Kerosene space heaters, coal and oil fuel combustion sources (0).	$80\ \mu g/m^3$ annual, $365\ \mu g/m^3$ for 24 h (EPA 1971).
Volatile organics Cooking; smoking; room deodorizers, cleaning sprays, paints, varnishes, solvents, and other organic products used in homes and offices; furnishings such as carpets, furniture, draperies; clothing; emissions from waste dumps (0).	None available.

infections also involve airborne transmission. The transmission occurs when the human respiratory tract emits liquid particles that evaporate to a particle size that can remain airborne for a period of time. Natural air currents or convective ventilation flows then transport the particles and deposit them in other human airways. The effect of reduced building ventilation on the incidence of infections is unknown.

A broad array of pollens, fungi, algae, actinomycetes, arthropod fragments, dusts, and pumices are confirmed airborne antigen sources that evoke adverse human responses; evidence is still emerging to implicate airborne bacteria, protozoa, and other groups in a similar manner. Although human exposure to airborne allergens recurs for varying periods of time, no reliable indoor or outdoor concentration data for allergens exist.

Carbon Monoxide

CO originates indoors primarily due to incomplete fuel combustion in unvented cooking and heating appliances and tobacco smoke. Vehicular emissions originating in attached or underground garages can also be an important source. CO is essentially nonreactive and in the absence of indoor sources, average indoor CO concentrations generally compare to outdoor concentrations. But if indoor sources are present, indoor levels can be much greater than those outdoors. Indoor levels can occasionally exceed the 8-h ambient standard, especially if significant indoor sources are present. Cases where the 1-h standard was exceeded have not been reported, except in instances where serious malfunctions in appliances may have occurred.

Formaldehyde

HCHO, formerly used in insulation, is a component in binders used in commercial wood products. Indoor sources of HCHO include particleboard, plywood, hardwood paneling, furniture, urea-formaldehyde foam insulation, tobacco smoke, and gas combustion. Some of the highest concentrations, exceeding 0.1 ppm, have been found in tightly constructed mobile homes where internal volumes are small compared with surface areas of HCHO-containing materials. HCHO emissions increase with increasing temperature and humidity.

Inhalable Particulate Matter

Until recently, measurements of particulate matter have centered on TSP, with essentially no size selection. Since the late 1970's, determinations have focused on IP of size 0 to 10 μm, with a coarse fraction of 3 to 10 μm and a fine fraction of under 3 μm. This fine fraction is often referred to as respirable particulate matter. Some scientists believe that it may be more important to stipulate considerations for particle size than to stipulate the exact size selection.

The fine and coarse fractions of IP have different sources and chemical compositions. Fine particles are mainly produced by coagulation of Aitken nuclei (<0.1 μm) and by vapor condensation onto these nuclei. Fine particles typically consist of sulfates, nitrates, ammonium salts, organics, and some metals such as lead (Pb) produced by various combustion processes and atmospheric transformations. Coarse particles are mainly produced by mechanical forces such as crushing and abrasion. Generally, these coarse particles consist of finely divided minerals such as oxides of silicon, iron, and aluminum; plant, animal, and insect fibers; tire particles; and sea salt.

Chemical analyses of IP suggest that indoor and outdoor compositions differ and that the building envelope acts as a barrier to outdoor sources. However, indoor IP mass may exceed outdoor levels, indicating that indoor sources such as smoking and reentrained dust are important determinants for indoor concentrations.

Metals and Other Inorganic Particulate Constituents

Metals found in the indoor environment include heavy elemental substances such as Pb, mercury (Hg), and arsenic (As). These substances are components of the particulate matter discussed earlier in this section. Evidence indicates that these metals have no significant indoor sources. In some older buildings where leaded paints were used, finely divided paint chips can become airborne through surface abrasion and other mechanisms. For young children in particular, direct ingestion is a greater hazard. In addition, smoking and the use of some pesticides contribute to indoor levels of heavy trace metals such as As and cadmium (Cd). Reentrainment is another possible indoor source when dust and particles enter a building either with the occupant or by infiltration.

Other inorganic constituents include sulfates and nitrates. Information on indoor generation of sulfates and nitrates is not available.

Nitrogen Dioxide

NO_2 sources generally are the same as those for CO, but NO_2 emissions result from high-temperature fuel combustion, whereas CO results from incomplete combustion. Further, NO_2 is a reactive gas. In the absence of indoor sources, indoor NO_2 levels are usually equal to or somewhat lower than outdoor concentrations. If indoor sources are present, indoor NO_2 concentrations can exceed outdoor levels; when the indoor source is turned off, the NO_2 concentration can decrease rapidly due to its chemical reactivity. Short-term (1-h or 24-h) indoor NO_2 concentrations in residences with indoor sources can also exceed the annual National Ambient Air Quality Standard (NAAQS) of 100 $\mu g/m^3$ (EPA 1971).

Ozone

In most cases, the source of indoor O_3 is ambient air. Exceptions include certain types of copying machines and air cleaners that work on electrostatic principles. O_3 decays very rapidly indoors. The half-life period for O_3, or the time required to reduce to one-half of the original concentration, is less than 30 min. Thus, sustained high indoor O_3 levels are seldom encountered.

Pesticides and Other Semivolatile Organics

Pesticides include a large group of commercially available toxic organic compounds used to control a variety of pests. Indoor sources of these substances include spray cans, pest strips and other coated surfaces, and contaminated fruits and flowers. Some of the pesticides commonly used in or near the indoor residential environment are chlordane, used to control carpenter ants and termites; dichlorvos, used in flea collars for dogs and cats; and carbamate, used in home insecticides. Limited data exist on indoor concentrations of pesticides.

Polychlorinated biphenyls (PCB) have excellent dielectric properties for use in electric transformers and capacitors. PCBs are no longer used in indoor applications, but large office buildings sometimes have PCB-containing transformers. Limited data exist on indoor PCB concentrations.

Polyaromatic Hydrocarbons

Polyaromatic hydrocarbons (PAH), also referred to as polycyclic organic material (POM) and polynuclear aromatics (PNA), represent a large family of complex organic substances that include known and suspected carcinogens. A combustion source emits PAH in a vaporous form that quickly condenses on suspended aerosols. Although benzo-a-pyrene (BaP) may not completely represent PAH exposures, BaP has often been measured as a surrogate indicator. PAH is derived from incomplete organic combustion in such processes as coke manufacture, asphalt production and use, and coal burning. The principal indoor sources of PAH are woodburning, smoking, and cooking.

Concentrations of these substances are in the nanogram-per-cubic-meter (ng/m^3) range; a great deal of debate has focused on the amount of total PAH missed by sampling only condensed PAH. However, exposure to vapor-phase PAH may not be as significant as PAH condensed onto particulate matter. Limited data on indoor concentrations are available.

Radon and Radon Progeny

Rn is a noble gas that has three naturally occurring radioactive isotopes (atomic masses of 219, 220, and 222) with half lives of 3.96 s, 55.6 s, and 3.82 d, respectively. Because of its longer half-life, radon-222 and its associated progeny, polonium-210, lead-214, bismuth-214, and polonium-214 (sometimes called radon daughters), are the principal sources of Rn exposure.*

Rn is spontaneously released from radium-containing geological materials. The gas may diffuse through air-filled pore spaces and fissures in the material or be transported by water and eventually enter the indoor airspace by bulk diffusion through foundation materials, diffusion through cracks, or entry through the water supply. Additionally, a building composed of radium-bearing material may itself be a source of Rn.

Rn progeny levels are related to Rn concentrations and both are determined by competing mechanisms of production and removal. However, progeny ions may be intercepted by indoor surfaces--this process is known as plateout--as well as

---

* Rn concentrations are stated in nanocuries per cubic meter (nCi/m^3), picocuries per liter (pCi/L), or Becquerels per cubic meter (Bq/m^3). A curie is defined as 3.7×10^{10} radioactive disintegrations per second. A Becquerel is defined as 1 disintegration per second. Thus, $1 nCi/m^3 = 1 pCi/L = 37 Bq/m^3$. Rn progeny activity is usually expressed in terms of working level (WL). One WL corresponds to any combination of Rn progeny in a liter of air that ultimately emits 1.3×10^5 megaelectronvolts (MeV) of alpha particle energy. In the ideal case, $1 nCi/m^3$ is equivalent to 0.01 WL. In realistic situations, $1 nCi/m^3$ may be equivalent to less than 0.01 WL for two reasons: progeny ions may attach to macro surfaces such as walls and thus escape detection, or aerosol removal processes may remove progeny already attached to particulate matter.

become attached to aerosols. Generally, between 50 and 95 percent of the progeny ions become attached to aerosols, some of which could leave the indoors due to air exchange.

Outdoor levels of Rn are generally on the order of 10^{-1} nCi/m^3, corresponding to 10^{-3} to 10^{-4} WL. Average indoor levels are estimated to be on the order of a few nanocuries per cubic meter (10^{-3} to 10^{-2} WL). Extreme cases, where levels exceeded 50 nCi/m^3 (about 10^{-1} WL), have been reported.

Sulfur Dioxide

Except for some kerosene space heaters and coal-burning stoves, indoor sources of SO_2 are rare. Like O_3, SO_2 also undergoes chemical transformation on indoor surfaces such as upholstery fabrics, draperies, and carpets, resulting in lower indoor concentrations. The half-life period for SO_2, however, is longer than for O_3. In the absence of indoor sources, indoor SO_2 concentrations in homes have generally been found to be lower than outdoor concentrations.

Volatile Organics

Many volatile organic vapor compounds are emitted indoors. These compounds are commonly found in modern building and decorating materials and in a variety of consumer products. Principal indoor sources of these compounds include solvents, furnishings, and other consumer products such as aerosols and coatings. Various indoor activities such as cooking, smoking, and arts and crafts also generate emissions of volatile organics. Concentrations of these pollutants vary widely from home to home, depending on source, strength, rate of ventilation, and other factors. Limited data on indoor and outdoor concentrations exist, but studies show that indoor concentrations exceed outdoor levels.

REFERENCES

Akland, G.G., Hartwell, T.D., Johnson, T.R., and Whitmore, R.W., Measuring Human Exposure to Carbon Monoxide in Washington, D.C., and Denver, Colorado, during Winter of 1982-1983, Environ. Sci. Technol., vol. 19, pp. 911-918, 1985.

American Society of Heating, Refrigerating, and Air Conditioning Engineers (ASHRAE) Ventilation for Acceptable Indoor Air Quality, ASHRAE Standard 62-1981, ASHRAE, Atlanta, 1981.

Andersen, I., Lundquist, G.R., and Molhave, L., Formaldehyde in the Atmosphere of Danish Homes, Ugeskr. Laeg., vol. 136, no. 38, pp. 2133-2139, 1974 [in Danish with English summary].

Benson, F.B., Henderson, J.J., and Caldwell, D.E., Indoor-Outdoor Air Pollution Relationships--A Literature Review, U.S. Environmental Protection Agency, Research Triangle Park, North Carolina, Publication AP-112, 1972.

Biersteker, K., DeGraaf, H., and Nass, Ch.A.G., Indoor Air Pollution in Rotterdam Homes, Int. J. Air Water Pollut., vol. 9, pp. 343-350, 1965.

Cole, J.T., and Zawacki, T.S., Emissions from Residential Gas-Fired Appliances, Gas Research Institute, Chicago, GRI-84/0164, 1985.

Cote, W.A., Wade, W.A. III, and Yocom, J.E., A Study of Indoor Air Quality, U.S. Environmental Protection Agency, Washington, D.C., EPA 650/4-74-042, 1974.

Drivas, P.J., Simmonds, P.G., and Shair, F.H., Experimental Characterization of Ventilation Systems in Buildings, Environ. Sci. Technol., vol. 6, pp. 609-614, 1972.

EPA. See U.S. Environmental Protection Agency.

Ferahian, R.H., Indoor Air Pollution--Some Canadian Experiences, Proceedings of the 3rd International Conference on Indoor Air Quality and Climate, Stockholm, vol. 1, p. 207, 1984.

Ferris, B.G., Speizer, F.L., Spengler, J.D., Effects of Sulfur Oxides and Respirable Particles on Human Health, Methodology, and Demography of Populations in Study, Am. Rev. Resp. Dis., vol. 120, p. 767, 1979.

Fugas, M., Assessment of Total Exposure to an Air Pollutant, Proceedings of the International Conference on Environmental Sensing and Assessment, Las Vegas, vol. 2, paper no. 38-5, 1975.

Janssen, J.E., The ASHRAE Ventilation Standard 62-1981: A Status Report, Proceedings of the 3rd International Conference on Indoor Air Quality and Climate, Stockholm, vol. 5, p. 199, 1984.

Leaderer, B.P., Zagraniski, R.T., Berwick, M., Stolwijk, J.A.J., and Qing-Shan, M., Residential Exposures to NO_2, SO_2, and HCHO Associated with Unvented Kerosene Space Heaters, Gas Appliances and Sidestream Tobacco Smoke, Proceedings of the 3rd International Conference on Indoor Air Quality and Climate, Stockholm, vol. 4, p. 151, 1984.

Lowder, W.M., George, A.C., Gogolak, C.V., Blay, A., Indoor Radon Daughter and Radiation Measurements in East Tennessee and Central Florida, Health and Safety Laboratory, U.S. Atomic Energy Commission, New York, HASL Technical Memorandum No. TM-71-8, 1971.

Maki, H.T., and Woods, J.E. Jr., Dynamic Behavior of Pollutants Generated by Indoor Combustion, Proceedings of the 3rd International Conference on Indoor Air Quality and Climate, Stockholm, vol. 5, p. 73, 1984.

Morey, P.R., Hodgson, M.J., Sorenson, W.G., Kollman, G.J., Rhodes, W.W., and Visvesvara, G.S., Environmental Studies in Moldy Office Buildings: Biological Agents, Sources and Preventive Measures, Ann. Am. Conf. Gov. Ind. Hyg., vol. 10, pp. 21-35, 1984.

Moschandreas, D.J., and Morse, S.S., Exposure Estimation and Mobility Patterns, Proceedings of the 72nd Annual Meeting of the Air Pollution Control Association, Pittsburgh, paper no. 79-14.4, 1979.

Nagda, N.L., and Koontz, M.D., Microenvironmental and Total Exposures to Carbon Monoxide for Three Population Subgroups, J. Air Pollut. Control Assoc., vol. 35, pp. 134-137, 1985.

Nagda, N.L., Koontz, M.D., and Rector, H.E., Energy Use, Infiltration, and Indoor Air Quality in Tight, Well-Insulated Residences, EPRI Report No. EA/EM-4117, Electric Power Research Institute, Palo Alto, California, June 1985.

National Research Council (NRC), Indoor Pollutants, National Academy Press, Washington, D.C., 1981.

National Research Council (NRC), An Assessment of Health Risk of Seven Pesticides Used for Termite Control, National Academy Press, Washington, D.C., 1982.

Sandberg, M., The Multi-Chamber Theory Reconsidered from the Viewpoint of Air Quality Studies, Build. Environ., vol. 19, pp. 221-233, 1984.

Shair, F.H., and Heitner, K.L., A Theoretical Model for Relating Indoor Pollutant Concentrations to Those Outside, Environ. Sci. Technol., vol. 8, pp. 444-451, 1974.

Skaret, E., Contaminant Removal Performance in Terms of Ventilation Effectiveness, Proceedings of the 3rd International Conference on Indoor Air Quality and Climate, Stockholm, vol. 5, p. 15, 1984.

Spedding, D.J., and Rowland, R.P., Sorption of Sulfur Dioxide by Indoor Substances--I. Wallpaper, J. Appl. Chem., vol. 20, pp. 143-146 (also see vol. 20, pp. 26-28 and vol. 21, pp. 68-70), 1970.

Traynor, G.W., Girman, J.R., Apte, M.G., and Dillworth, J.F., Indoor Air Pollution Due to Emissions from Unvented Gas-Fired Space Heaters, J. Air Pollut. Control Assoc., vol. 35, p. 231, 1985.

Turk, A., Measurements of Odorous Vapors in Test Chambers: Theoretical, ASHRAE J., vol. 5, no. 10, pp. 55-58, 1963.

U.S. Environmental Protection Agency, National Primary and Secondary Ambient Air Quality Standard (40 CFR 50), Federal Register, vol. 36, p. 22384, 1971.

U.S. Environmental Protection Agency, National Primary and Secondary Ambient Air Quality Standards for Carbon Monoxide (40 CFR 50.8), Federal Register, vol. 40, p. 7043, 1975.

U.S. Environmental Protection Agency, National Primary and Secondary Ambient Air Quality Standards for Lead (40 CFR 50.12), Federal Register, vol. 43, p. 46258, 1978.

U.S. Environmental Protection Agency, Proceedings of the Symposium on the Development and Usage of Personal Monitors for Exposure and Health Effects Studies, Research Triangle Park, North Carolina, EPA-600/9-79-032, 518 pages, 1979a.

U.S. Environmental Protection Agency, National Primary and Secondary Ambient Air Quality Standards for Ozone (40 CFR 50.9), Federal Register, vol. 44, p. 8220, 1979b.

U.S. Environmental Protection Agency, Review of the National Ambient Air Quality Standards for Particulate Matter: Revised Draft Staff Paper, Research Triangle Park, North Carolina, 1981.

Ware, J.H., Dockery, D.W., Spiro, A. III, Speizer, F.E., and Ferris, B.G. Jr., Passive Smoking, Gas Cooking, and Respiratory Health of Children Living in Six Cities, Am. Rev. Respir. Dis., vol. 129, pp. 366-374, 1984.

Yocom, J.E., Clink, W.L., Cote, W.A., Indoor/Outdoor Air Quality Relationships, J. Air Pollut. Control Assoc., vol. 21, p. 251, 1971.

Chapter 3
FACTORS AFFECTING INDOOR AIR QUALITY

Indoor air quality depends on many factors. An indoor source that emits pollutants can increase the pollutant concentrations and degrade indoor air quality.
However, bringing in outdoor air, if of better quality, can improve the overall
indoor air quality. Similarly, the rate at which outdoor air is exchanged with
indoor air, the mixing of indoor air, and characteristics of the pollutants all
influence indoor air quality. To further complicate the issue, two or more of
these factors can act simultaneously; thus, the influence of each on indoor air
quality may not be easily discernible.

A clear understanding of how these and other factors influence indoor air quality
will allow for the development of an effective and efficient design for monitoring indoor air quality. An effective design enables sufficient information to be
gathered to meet the objectives that prompted the monitoring. An efficient
design enables the researcher to collect the information at lower resource or
cost levels or to collect more information for the same cost.

The first half of this chapter uses illustrations to explain how each factor
individually affects indoor air quality. The mathematical basis for these illustrations is provided in the second half.

HOW IS INDOOR AIR QUALITY AFFECTED?

Outdoor Concentrations

Outdoor concentrations affect indoor concentrations. In Figure 1, outdoor and
indoor concentrations are plotted for the same time period. As the outdoor concentration of carbon monoxide (CO) rises from 1 to 9 mg/m^3, primarily due to
local vehicular traffic, the indoor concentration of CO also increases; however,
the indoor concentration rises at a slower rate than the outdoor. In addition,
the indoor CO concentration peaks somewhat later than the outdoor concentration.
Similarly, when the outdoor concentration falls, beginning at 9 a.m., the indoor
concentration follows suit, but again at a slower rate. Thus, the building envelope has a dampening or shielding effect on the indoor peak concentration. This
effect is only temporary, because if the outdoor concentration stays constant at
some concentration value, the indoor concentration will eventually equal that of
the outdoor concentration.

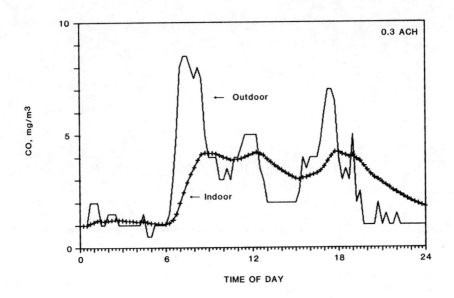

FIGURE 1. Effect of outdoor concentrations on indoor concentrations.

Indoor Sources

Activation of indoor sources increases indoor concentrations. The effect of indoor sources is illustrated in Figure 2. An unvented combustion appliance--gas kitchen range--was operated for two different time periods (60 min in the morning and 30 min in the afternoon). The increase in indoor concentration was greater in the morning when the source was operated for a longer time. Although formaldehyde (HCHO) sources are not "operated" in the usual sense, HCHO emanations increase with rising indoor temperature and humidity. Thus, the variation in pollutant emissions, either due to operational changes or changes in environmental conditions, must be considered in evaluating the impact of a source on indoor air quality.

Air Exchange Rate

Air exchange rates govern the rate at which indoor concentrations rise or fall due to changes in outdoor concentrations. The air exchange rate is the rate at which indoor air is exchanged with outdoor air. The rate is expressed as the volume of air exchanged per unit time. The volume of air is commonly expressed in terms of the volume of a structure and 1 h is used as the time unit. So, for a house with a 200-m^3 volume, where 50 m^3 of outdoor air is entering per hour to replace the same amount of indoor air, the air exchange rate is 0.25 air changes per hour (ACH).

In Figure 1, the air exchange rate is 0.3 ACH. When the air exchange rate is 0.9 ACH (Figure 3), the indoor concentration follows the outdoor concentration more rapidly and more closely. With a lower exchange rate (0.1 ACH), the increases and decreases in indoor concentrations are much slower (Figure 4). The outdoor concentrations in Figures 3 and 4 are the same as those in Figure 1.

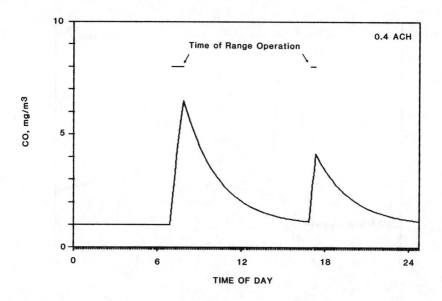

FIGURE 2. Effect of duration of indoor source operation on indoor concentrations.

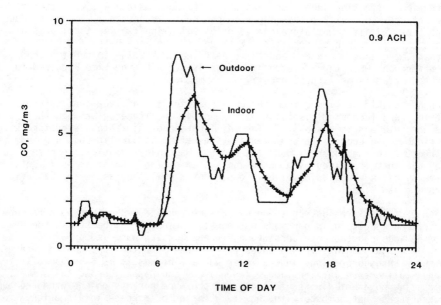

FIGURE 3. Effect of outdoor concentrations on indoor concentrations at 0.9 ACH.

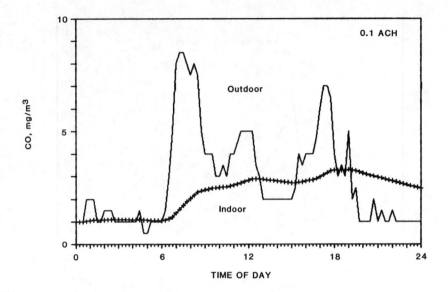

FIGURE 4. Effect of outdoor concentrations on indoor concentrations at 0.1 ACH.

The rate of air exchange also determines the amount of time required for the indoor pollutant concentration to decrease. As in the case of outdoor pollutant concentrations affecting indoor concentrations (Figures 3 and 4), the rate of air exchange affects how quickly the indoor concentration decreases. Figure 5 shows the impact of two different air exchange rates following two identical episodes (60 min each) of gas range operation. With an air exchange rate of 0.8 ACH in the morning, the concentration reduces to half the peak value in about an hour; in the afternoon, when the air exchange range is 0.2 ACH, the time required for a similar reduction is more than three times as much.

Air exchange should be viewed from three perspectives: infiltration, natural ventilation, and mechanical ventilation. A pressure difference between the indoors and outdoors is responsible for infiltration of air into a structure with closed windows and doors. Natural ventilation refers to air flowing into and out of a structure through open windows and doors. Mechanical ventilation refers to air exchange that is driven by a motorized system or fan. Local or spot ventilation refers to the ventilation of only a part of a building, as is the case with a bathroom exhaust fan.

In the case of infiltration, pressure differences are caused either by a temperature difference between indoors and outdoors, by wind, or by both. In the winter, the indoor-outdoor temperature difference causes warm indoor air to rise and leave through openings in the upper part of the structure and cooler outdoor air to enter through openings in the lower part. The result is a "stack effect" that tends to increase infiltration. The reverse can happen during the summer. Wind impacting on a building surface creates pressure pockets on the windward and leeward surfaces that combine with the stack force to increase infiltration.

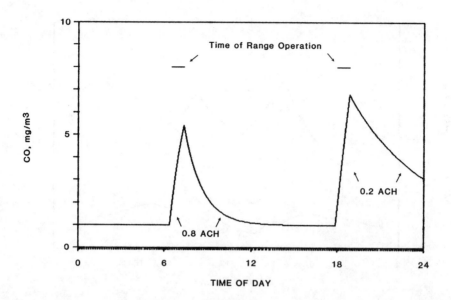

FIGURE 5. **Effect of air exchange and indoor source operation on indoor concentrations.**

Additionally, wood stoves and fireplaces can interact with stack and wind forces to increase air infiltration under conditions of strong stack effect or chimney draft. However, when the chimney draft is weak, "back drafts" can develop to pull combustion products into the living space. Some types of modern wood stoves derive combustion air from a special vent to preclude such occurrences; back drafts can still develop during initial lighting and subsequent reloading.

Although sharply defined changes in air exchange rates such as those shown in Figure 5 can be a result of mechanical ventilation only, air infiltration rates can vary from hour to hour in any particular day and can vary substantially over different seasons. Figure 6 shows measured hourly infiltration rates that vary from <0.1 to >0.7 ACH over two seasons for the same building. These variations are due to changes in the prevailing wind and stack forces.

Volume

Indoor concentration is also dependent on the volume that is available for the pollutants to disperse. The volume for pollutant dispersal can be the entire volume of the structure. The same amount of source emissions released into a smaller structure with 100 m^2 (approximately 1 100 ft^2) floor area and a larger structure with 200 m^2 would result in different concentrations. Figure 7 shows CO concentrations due to a convective kerosene space heater that was operated for supplemental heat. The concentrations in a 100-m^2 structure are approximately twice as high as those in a 200-m^2 structure.

Under certain conditions, not all of the indoor volume is equally available for the pollutant to disperse. This situation is easy to recognize if the source is

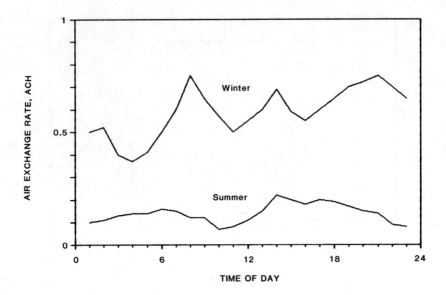

FIGURE 6. Hourly air infiltration rates for a summer and a winter day (summer--windspeeds, 0 to 2 m/s; outdoor temperatures, 18 to 31 °C; indoor temperature, 25 °C; winter--windspeeds, 0.3 to 4 m/s; outdoor temperatures, −10 to 0 °C; indoor temperature, 18 °C). Adapted from Nagda et al. 1985.

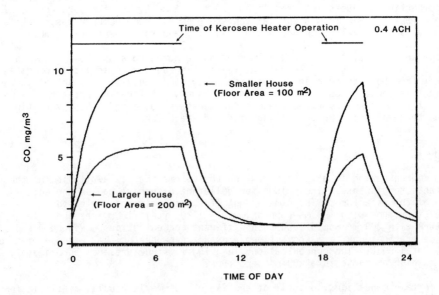

FIGURE 7. Effect of the volume of the structure on indoor pollutant concentrations.

located in a room and the door to that room is closed. However, there are times
when dispersion can be blocked by invisible barriers. For example, if an indoor
combustion source is located on an upper floor of a building, the pollutants
would disperse mainly on the upper floor and those floors above that floor. This
tendency is due to thermal buoyancy, which restricts emissions from dispersing to
a lower floor. Similarly, if a source for a contaminant is on a lower floor, its
concentration elsewhere in the structure depends on air circulation from that
floor to other floors.

Figure 8 shows radon (Rn) concentrations for an upper floor and lower floor of a
two-story house. Rn enters the structure from the soil underneath the structure
into the lower floor. Thus, without forced mixing, Rn concentrations are higher
on the lower floor than on the upper floor. However, when a circulation fan for
either a heating or air conditioning system is operating (during hours 8 to 24 and
36 to 48 in Figure 8), the concentration of the upper floor tends to rise because
it is mixed with radon-rich air from the lower floor. Similarly, with the mixing,
the downstairs concentration tends to decrease. When the circulation fan is off,
Rn tends to accumulate on the lower floor, thereby increasing the difference in
concentrations between the two floors.

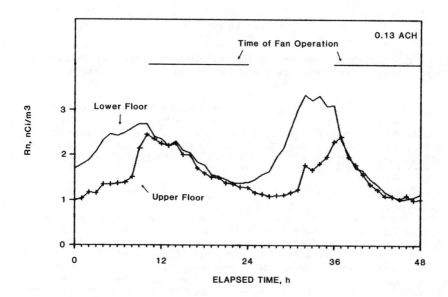

FIGURE 8. Effect of circulation fan on indoor radon concentrations.

Pollutant Characteristics

The characteristics of a pollutant are important in determining indoor concentrations. CO and nitrogen dioxide (NO_2) that result from the same combustion
source behave differently even though they are released at the same time and are
under the same conditions of air exchange rates (Figure 9). Reduction in NO_2
concentration is due not only to indoor air leaving the structure through air
exchange, as in the case of CO, but also to a chemical reaction that NO_2 undergoes, probably with various available indoor surfaces. Thus, its concentration
decreases much more rapidly than CO.

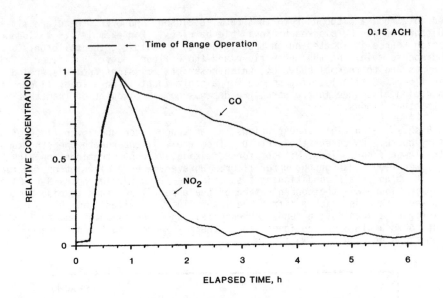

FIGURE 9. Comparison of decrease in NO₂ and CO concentrations following a gas range operation (gas range operated for 40 min during the first hour).

The impact of air exchange on HCHO is less than what can be expected for other pollutants such as CO. An increase in air exchange by a factor of two may only reduce pollutant concentration by 35 percent. This limited reduction occurs because as the air exchange rate increases, the rate at which HCHO emanates also increases. Thus, pollutants or, in this case, pollutant-source characteristics have an important influence on indoor air quality.

Pollutant Removal

Ventilation. Mechanical ventilation can reduce concentrations of indoor-generated pollutants. The impact of mechanical ventilation is similar to that of air infiltration, but the location of exhaust can play an important role in determining indoor air quality. For example, an exhaust fan located directly adjacent to an indoor source will have a greater impact than one that ventilates the entire structure. Figure 10 shows three cases of gas range operation: (1) with no exhaust, such that the air exchange rate is composed solely of air infiltration, (2) with an increase in the air exchange rate by mechanical ventilation for the entire structure, and (3) with operation of a range-hood fan that is vented outdoors during operation of the gas range. In Figure 10, where it is assumed that infiltration is constant at 0.3 ACH, mechanical ventilation adds an additional 0.2 ACH; thus, the effective air exchange with mechanical ventilation is 0.5 ACH. When the range fan is operating, it adds 0.2 ACH to infiltration. Thus, the effective air exchange with the range fan during the range operation is also 0.5 ACH. For the remainder of the time, air exchange is solely due to air infiltration, which is 0.3 ACH.

For case 2, the rise in concentration is somewhat slower and the decrease in concentration is much faster than for case 1; for case 3, the rise in concentration is much less than for the other two cases. With the range fan, some of the combustion contaminants are removed before they have a chance to mix with the interior air volume; consequently, the rise in indoor concentration is limited. Thus, even though both the mechanical ventilation and range fan cases have the same total air exchange during the period of range operation, the range fan is much more effective in reducing indoor concentrations. An additional advantage is that the range fan operates only during a limited time period whereas the mechanical ventilation operates continuously; thus, mechanical ventilation is likely to result in greater energy consumption than the range fan.

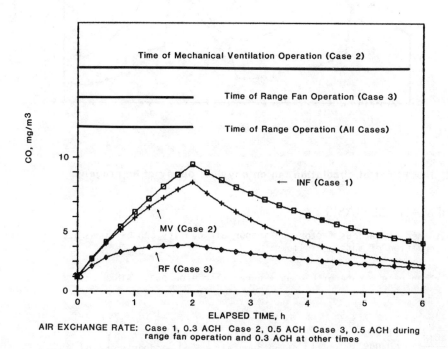

FIGURE 10. Comparison of effectiveness of mechanical ventilation (MV) and range fan (RF) operation against infiltration (INF).

Cleaning devices and other removal mechanisms. Air cleaners can remove indoor pollutants without resorting to large air exchange rates. A number of air cleaning devices are available, ranging from small units sized to treat an individual room to larger units designed to handle an entire building.

In certain instances, a property of the pollutant can aid in reducing indoor concentrations. In the case of Rn and Rn progeny, for example, the enhanced plateout and filtration effects of a central air handling system can substantially reduce working levels without increasing air exchange rates. As shown in Figure 11, when the central fan system is off, the correspondence between Rn and Rn progeny approaches 0.01 WL/nCi/m^3. With the fan running, a substantial fraction of the progeny is removed by enhanced plateout and filtration. The progeny levels are reduced to approximately 0.005 WL/nCi/m^3, a reduction factor of nearly 50 percent, without affecting the air exchange rate. The Rn concentration, however, is unaffected, and remains in the form of its normal diurnal curve.

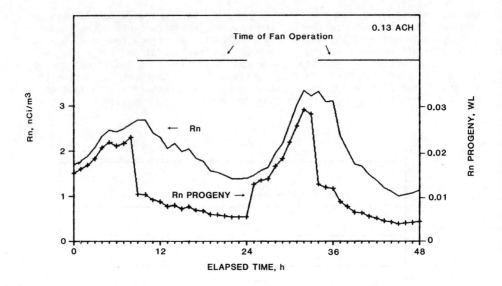

FIGURE 11. Effect of circulation fan on physical decay of Rn progeny.

MATHEMATICAL BASIS

The influence of various factors on indoor air quality can be summarized based on the conservation of mass as follows:

accumulation rate = rate of [input + generation - output - sink]

or

$$\frac{dx}{dt} = \frac{VdC_{in}}{dt} = \begin{array}{l}\text{rate of} \\ \text{change in} \\ \text{mass due to}\end{array} \left[\left(\begin{array}{c}\text{infiltration} \\ \text{of outdoor air}\end{array}\right) + \left(\begin{array}{c}\text{generation} \\ \text{indoors}\end{array}\right)\right.$$

$$\left.- \left(\begin{array}{c}\text{exfiltration} \\ \text{of indoor air}\end{array}\right) - \left(\begin{array}{c}\text{indoor removal} \\ \text{of pollutants}\end{array}\right)\right] \qquad (1)$$

where

V is the indoor volume

C_{in} is the indoor concentration.

Lidwell and Lovelock (1946) were among the first to use the mass balance approach. Some years later, Turk (1963) presented general equations to describe specific cases of transient and steady-state conditions in a single-chamber model. Refinements to this basic approach are summarized in more recent publications (National Research Council 1981; Nagda et al. 1985).

The four terms on the right-hand side of equation 1 and the term, V, for volume, are explained in the following discussion. Substituting proper expressions for each of the four terms on the right-hand side of equation 1 yields a differential equation that can be solved for specific situations.

Infiltration of Pollutants from Outdoors

The amount of a pollutant that infiltrates indoors is a product of two factors: (1) volume rate of air exchange (νV, where ν is the air exchange rate measured in air changes per hour) between outdoors and indoors through the building envelope and (2) outdoor pollutant concentrations (C_{out}). Additionally, when outdoor air enters a structure, a certain fraction of the outdoor pollutant concentration, F_B, may be intercepted by the cracks and crevices in the building envelope, resulting in a filtration or scrubbing effect. Thus, the infiltration of pollutants from outdoor air over an incremental time period, dt, can be expressed as $(1 - F_B)\nu V C_{out} dt$.

Indoor Generation

A variety of indoor appliances and materials generates certain pollutants. When a source is constantly producing these contaminants for a time period, dt, the indoor generation can be expressed as Sdt, where S is the rate of indoor emission.

Exfiltration of Indoor Air

Exfiltration, like infiltration, is the product of volume rate of air exchange (νV) and the concentration of air (C_{exit}) leaving the structure. In cases where the indoor space can be assumed to be well mixed, C_{exit} is the same as C_{in}. Thus, the exfiltration term can be expressed as $\nu V C_{in} dt$.

Indoor Pollutant Removal or Decay

Certain pollutants such as NO_2, ozone, and sulfur dioxide decrease in concentration due to chemical reactions and/or adsorption of the contaminants, particularly on indoor surfaces. Radioactive decay is a fundamental property for Rn and Rn progeny. However, Rn progeny are also subject to physical and chemical removal, complicating the understanding of the overall decay process for Rn. The decay can be expressed as λdt, where λ is the overall rate of decay. When more than one decay process is active, each decay term (λ^i) can be considered separately such that $\lambda = \Sigma \lambda^i$. Indoor pollutants can also be removed through air cleaning devices. This process, which depends on the volume of air going through a cleaning device and on the efficiency of the device, can be expressed as $qFC_{in} dt$, where q is the volume flow rate and F is the fraction removed by cleaning devices.

Volume

The indoor volume of a structure has a direct influence on the indoor concentration when a pollutant is generated indoors. However, effective volume (cV)--the volume that is actually involved in pollutant dispersal--depends on the degree of air circulation. The value of c is unity when the whole indoor volume is involved, as in the case when a central circulation fan is constantly operating. When there is no forced mixing, the degree of circulation is partly dependent on thermal gradients within and between various indoor spaces.

Generalized Mass Balance Equation

Considering the four terms in equation 1 and assuming the effective volume is cV instead of the total indoor volume V, a generalized mass balance equation for indoor concentration under well-mixed conditions would be:

$$cVdC_{in} = (1 - F_B) \nu cVC_{out}dt + Sdt - \nu cVC_{in}dt - \lambda dt - qFC_{in}dt$$

or

$$\frac{dC_{in}}{dt} = (1 - F_B) \nu C_{out} + \frac{S}{cV} - \nu C_{in} - \frac{\lambda}{cV} - \frac{qFC_{in}}{cV} \tag{2}$$

where

C_{in} = indoor concentration (units: mass/volume)

F_B = fraction of outdoor concentration intercepted by the building envelope (dimensionless fraction)

ν = air exchange rate (1/time)

C_{out} = outdoor concentration (mass/volume)

S = indoor generation rate (mass/time)

cV = effective indoor volume where c is a dimensionless fraction (volume)

λ = decay rate (mass/time)

q = flow rate through air cleaning device (volume/time)

F = efficiency of the air cleaning device (dimensionless fraction).

To account for imperfect mixing of the interior, a mixing factor is introduced. As the variable c modifies indoor volume to yield effective volume, the mixing factor modifies the air exchange to give an effective air exchange rate for a pollutant.

The mixing factor, m, a dimensionless number, is defined as the ratio of actual residence time of the pollutant indoors to the residence time derived directly from the air exchange rate (Constance 1970; Drivas et al. 1972; Esmen 1978). It can also be expressed as the ratio of the concentration of the "exit stream" to the indoor concentration. The exit stream concentration is difficult to quantify, especially when the indoor mass is leaving the structure as a result of uncontrolled exfiltration rather than leaving the structure at specific points, such as from a range-hood exhaust.

The effective air exfiltration rate for a pollutant is the product of m and the air exchange rate, or equal to $m\nu$. Thus, in the absence of ideal mixing conditions, equation 2 can be modified to:

$$\frac{dC_{in}}{dt} = (1 - F_B) \nu C_{out} + \frac{S}{cV} - m\nu C_{in} - \frac{\lambda}{cV} - \frac{qFC_{in}}{cV} . \tag{3}$$

For each pollutant and prevailing conditions, appropriate substitutions can be made in equation 3. For example, for the simplest case involving a chemically inert gas such as CO, λ can be assumed to be 0. If there are no indoor sources and no cleaning devices, $S = 0$ and $F = 0$.

Similarly, if it is assumed that $F_B = 0$ and that perfect mixing exists, then equation 3 reduces to:

$$\frac{dC_{in}}{dt} = \nu C_{out} - \nu C_{in} \, . \tag{4}$$

To solve this differential equation, consider the boundary conditions:

$C_{in, \, initial} = C_{in,o}$

$C_{in, \, final} = C_{out}$ where C_{out} is assumed to remain constant.

Thus, the solution of equation 4 for any later time, t, is:

$$C_{in,t} = C_{out} + (C_{in,o} - C_{out}) \, e^{-\nu \Delta t} \tag{5}$$

where $\Delta t = t - t_o$. This equation can be applied to subsequent Δt intervals by substituting the calculated concentration for the first interval as the initial concentration for the next interval in place of $C_{in,o}$ in equation 5. These calculations are then repeated for the total time period of interest. In this example C_{out} was deliberately kept constant, but time-varying values of C_{out}, i.e., $C_{out,ti}$ can be used for a more realistic estimation of indoor concentrations. Time-varying values of C_{out} were used for the case shown in Figure 1. Some useful solutions of the generalized mass balance equation used to prepare some of the figures in this chapter are given in Table 2.

The preceding discussion of the mass balance model has been restricted to consider a single chamber, as in the case of one room or a well-mixed house. In many cases, however, it is common to consider two or more chambers to properly simulate physical conditions--such as in the case of a building with multiple floors--especially if mixing of air among the floors is restricted. Theoretical models have been developed (Sinden 1978; Sandberg 1984) and some experimental work has been conducted to use such models from the perspective of indoor air quality (Hernandez and Ring 1982) as well as from strictly ventilation perspectives (Skaret and Mathison 1983). But the area of multizone modeling still remains largely unexplored.

Terms in the mass balance equation are listed in Table 3. Many variables influence these terms; some of these are listed in Table 4. The list is not exhaustive but can be used as a starting point for choosing parameters that need to be measured. Techniques and equipment for conducting measurements are described in the next chapter.

TABLE 2. Some Solutions to Mass Balance Equations

Illustration/conditions	Solution

Figure 1

S = No indoor source

ν = 0.3 ACH for all t

t_i = initial time of calculation where $t_i < t$

$\Delta t = t - t_i$

$C_{in,o}$ = 1 mg/m^3

C_{out,t_i} = displayed

F_B = 0

c and m = 1

λ = 0

q = 0

V = 200 m^3

$$C_{in,t} = C_{out,t_i} + (C_{in,t_i} - C_{out,t_i})\, e^{-\nu\Delta t}$$

Figure 2

S = 800 mg/h

ν = 0.4 ACH for all t

$C_{in,o}$ = 1 mg/m^3

C_{out} = 1 mg/m^3 for all t

V = 120 m^3

$\Delta t = t - t_i$, where

t_i = initial time of calculation where $t_i < t$

t_1 = time source turned on (two cases: t_1 = 0700 h and t_1 = 1700 h)

t_2 = time source turned off (two cases: t_2 = 0800 h and t_2 = 1730 h)

For time periods given by $t_i < t_1$ or $t_i \geq t_2$, the solution is the same as that for Figure 1.

For time periods given by $t_1 \leq t_i < t < t_2$ when source is operating

$$C_{in,t} = C_{out} + \frac{S}{\nu V} + \left(C_{in,t_i} - C_{out} - \frac{S}{\nu V} \right) e^{-\nu\Delta t}$$

Figures 3 and 4

ν = 0.9 ACH (Figure 3)

ν = 0.1 ACH (Figure 4)

All other parameters are the same as those for Figure 1

Same solution as that for Figure 1

(Continued)

TABLE 2. Some Solutions to Mass Balance Equations (Continued)

Illustration/conditions	Solution

Figure 5

t_1 = 0630 and 1800 h

t_2 = 0730 and 1900 h

ν = 0.8 ACH (until 1800 h)

ν = 0.2 ACH (from 1800 to 2400 h)

All other paramaters are the same as those for Figure 2

Same solution as that for Figure 2 with source operating

Figure 7

S = 210 mg/h

ν = 0.4 ACH for all t

$C_{in,o}$ = 1 mg/m^3

C_{out} = 1 mg/m^3 for all t

t_1 = 0000 and 1800 h

t_2 = 0800 and 2100 h

V = 244 m^3 (smaller house)

V = 488 m^3 (larger house)

Same solution as that for Figure 2 with source operating

Figure 10

S = 1000 mg/h

t_1 = 0000 h

t_2 = 0200 h

$C_{in,o}$ = 1 mg/m^3

C_{out} = 1 mg/m^3 for all t

V = 200 m^3

ν_{INF} = 0.3 ACH for all t

ν_{MV} = 0.2 ACH for all t

For $t_1 \leq t_i < t < t_2$ when range is operating

Case 1: Only infiltration (INF) and

Case 2: Infiltration and mechanical ventilation (MV):

$$C_{in,t} = C_{out} + \frac{S}{\nu V} + \left(C_{in,t_i} - C_{out} - \frac{S}{\nu V} \right) e^{-\nu \Delta t}$$

where $\nu = \nu_{INF}$ for Case 1

$\nu = \nu_{INF} + \nu_{MV}$ for Case 2

(Continued)

TABLE 2. Some Solutions to Mass Balance Equations (Concluded)

Illustration/conditions	Solution

Figure 10 (Concluded)

ν_{RF} = 0.2 ACH (range fan operating only when source was operating)

Case 3: <u>With range exhaust fan (RF):</u>

$$C_{in,t} = \frac{1}{\nu + m\nu_{RF}} \left[\left((\nu + \nu_{RF})C_{out} + \frac{S}{V} \right) \left(1 - e^{-(\nu + m\nu_{RF})\Delta t} \right) \right.$$
$$\left. + (\nu + \nu_{RF}) \, C_{in,t_i} \, e^{-(\nu + m\nu_{RF})\Delta t} \right]$$

For $t \geq t_2$:

$$C_{in,t} = C_{out} + \left(C_{in,t_i} - C_{out} \right) e^{-\nu\Delta t}$$

where $\nu = \nu_{INF}$ for Cases 1 and 3

$\nu = \nu_{INF} + \nu_{MV}$ for Case 2

Figure 11

WL_{in} = working level (Rn progeny concentration)

C_{in}^0 = displayed (indoor Rn concentration)

ν = 0.13 ACH

λ^1 = 13.6 h^{-1}

λ^2 = 1.5 h^{-1}

λ^3 = 2.1 h^{-1}

C_1 = 0.00103 WL/(nCi/m^3)

C_2 = 0.00507 WL/(nCi/m^3)

C_3 = 0.00373 WL/(nCi/m^3)

K_R = 1.5 h^{-1} without fan,

3.0 h^{-1} with fan

$$WL_{in} = C_1 \left(\frac{\lambda^1 C_{in}^0}{K_1} \right) + C_2 \left(\frac{\lambda^1 \lambda^2 C_{in}^0}{K_1 K_2} \right) + C_3 \left(\frac{\lambda^1 \lambda^2 \lambda^3 C_{in}^0}{K_1 K_2 K_3} \right)$$

$K_i = \nu + \lambda^i + K_R$ for i = 1, 2, and 3

where i = 1, 2, and 3, and corresponds to 1st, 2d, and 3d generations of progeny

C_1, C_2, and C_3 are contributions to working level for the respective progeny

TABLE 3. Terms in Mass Balance Equation

Output term of the mass balance equation	Input terms
Indoor air quality, C_{in}	Air exchange, ν Outdoor air quality, C_{out} Source rate, S Effective volume, cV Decay rate, λ Building filtration factor, F_B Other removal mechanisms Mixing factor, m

TABLE 4. Variables that Influence Input Terms in Mass Balance Equation

Input terms	Variables that influence input terms
Air exchange, ν	
Infiltration, ν_{INF}	Indoor-outdoor temperature differential Windspeed, wind direction, wind barriers Type of structure Tightness of structure

Mechanical ventilation, ν_{MV}	Building ventilation system
	  - capacity
	  - mode of operation
	  - patterns of use
	Local exhaust or spot ventilation
	  - location
	  - capacity of fan
	  - patterns of use
	Air-to-air heat exchanger

Natural ventilation, ν_{NV}	Ambient conditions
	Windows or doors
	  - size
	  - type
	  - location
	  - patterns of use

Outdoor air quality, C_{out}	Geography
	  - spatial relation to outdoor sources
	  - terrain
	Geology
	Season
	Time of day

(Continued)

TABLE 4. Variables that Influence Input Terms in Mass Balance Equation (Concluded)

Input terms	Variables that influence input terms

Source rate, S	Appliance/furnace
	  - type
	  - size, capacity
	  - mode of operation
	  - maintenance
	  - patterns of use
	Building materials
	Type of foundation
	Consumer products

Effective volume, cV	Circulation
	  - natural
	    · infiltration
	    · thermal gradients
	    · people movement
	  - mechanical
	    · circulation fan (fan-forced
	      heating/cooling)
	    · location of supply and
	      return vents
	    · spot circulation (table
	      fan, ceiling fan)

Decay rate, λ	Surface
	  - type
	  - area or number of active sites
	  - surface-to-volume ratio
	Relative humidity
	Temperature
	Radioactive half-life

Building filtration factor, F_B	Building materials Building construction Type of pollutant

Other removal mechanisms	Filter
	  - type
	  - maintenance
	Air cleaning device
	  - type
	  - capacity
	  - patterns of use

Mixing factor, m	Local exhaust

REFERENCES

Constance, J.D., Mixing Factor is Guide to Ventilation, Power, vol. 114, no. 2, pp. 56-57, 1970.

Drivas, P.J., Simmonds, P.A., Shair, F.H., Experimentation Characterization of Ventilation Systems in Buildings, Environ. Sci. Technol., vol. 6, pp. 609-614, 1972.

Esmen, N.A., Characterization of Contaminant Concentrations in Enclosed Spaces, Environ. Sci. Technol., vol. 12, pp. 337-339, 1978.

Hernandez, T.L., and Ring, J.W., Indoor Radon Source Fluxes: Experimental Tests of a Two Chamber Model, Environ. Int., vol. 8, pp. 45-57, 1982.

Lidwell, O.M., and Lovelock, J.E., Some Methods of Monitoring Ventilation, J. Hyg., vol. 44, pp. 326-332, 1946.

Nagda, N.L., Koontz, M.D., and Rector, H.E., Energy Use, Infiltration, and Indoor Air Quality in Tight, Well-Insulated Residences, Electric Power Research Institute, EPRI EA/EM 4117, Palo Alto, 1985.

National Research Council, Committee on Indoor Pollutants, Indoor Pollutants, Washington, D.C.: National Academy Press, 1981.

Sandberg, M., The Multi-Chamber Theory Reconsidered from the Viewpoint of Air Quality Studies, Build. Env., vol. 19, no. 4, pp. 221-233, 1984.

Sinden, F.W., Multichamber Theory of Infiltration, Build. Environ., vol. 13, pp. 21-28, 1978.

Skaret, E., and Mathison, H.M., Ventilation Efficiency--A Guide to Efficient Ventilation, ASHRAE Trans., vol. 89, DC-83-02 No. 2, 1983.

Turk, A., Measurements of Odorous Vapors in Test Chambers: Theoretical, ASHRAE J., vol. 5, no. 10, pp. 53-58, 1963.

Chapter 4
MEASUREMENTS

In indoor air quality monitoring, the word "measurement" should be used in a broad sense; it includes measurements conducted through the use of physical or chemical methods for sampling and analyzing pollutant concentrations, air exchange rates, or environmental parameters. Additionally, questionnaires and other types of survey instruments are used in indoor air quality studies to obtain measurements of related and equally important information such as characteristics of the building; heating, ventilating, and air conditioning (HVAC) systems; appliances; and occupant activity patterns. Much of this chapter describes physical devices and chemical methods, reflecting extensive and systematic efforts in development and testing. Although survey instruments have not been as extensively developed and tested, their importance is recognized. Fundamental points that should be considered in the development and use of survey instruments are described in the final section of this chapter.

DEFINITIONS FOR MONITORING INSTRUMENTS AND METHODS

For convenience, two types of monitoring systems are defined: those that are commercially available as preassembled devices and those that are assembled from components. Devices of the first type are referred to in this chapter as "instruments" (for example, an instrument to measure carbon monoxide concentrations). Such commercially available instrumentation offers obvious benefits; the measurement techniques are based on proven technical principles and the time or effort needed to assemble and to test measurement systems can be minimal. For some applications, however, commercial instrumentation may be unavailable or too expensive.

Devices of the second type are referred to as "methods" because they require assembly of various components, possibly from different suppliers, to construct a system that conforms to a standard or an accepted technique. Measurements of organic vapors, for example, may be carried out by a number of methods, most of which involve a sorbent trap and air pump for sample collection followed by gas chromatography using various detection systems such as mass spectrometry, electron capture, or flame ionization.

Sampling mobility, operating characteristics, and output characteristics are basic considerations when selecting an instrument or a method. There are three classes of sampling mobility:

● Personal--the unit may be conveniently carried or worn

● Portable--the unit may be hand-carried from one place to another during sampling, but does not offer the convenience of a personal device

● Stationary--the unit must operate from a fixed location.

Obviously, either a personal or portable measurement system can be used in a stationary mode. Because personal monitors are not available for many pollutants, portable instruments or methods are often the only recourse for personal monitoring. Such instruments are often more convenient to use in indoor environments and frequently are less expensive than the equivalent stationary instruments.

Within each class of mobility are two categories of operating characteristics:

● Active--a power source is required to draw sample air to a sensor or collector

● Passive--no power source is required; sample acquisition relies on diffusion.

Finally, within each mobility and operating class, the output characteristics fall into the following categories:

● Analyzer--the unit produces an instantaneous (or nearly instantaneous) signal that corresponds to the pollutant concentration or the parameter that is being measured

● Collector--The sample is collected by the unit and subsequently is analyzed and quantitated in a laboratory.

Analyzers produce time-series information or output and are useful in determining peak concentrations. The use of analyzers to obtain time-series output is often referred to as continuous monitoring. Such output from analyzers can be integrated over time to estimate time-weighted average concentrations. Data obtained from collectors are limited to time-weighted average concentrations. Information on peak concentrations is not available unless averaging periods are very short and suitable analytical methods are available to quantitate relatively lower mass collected over a short time period.

Pollutants of interest in indoor air quality monitoring include carbon monoxide (CO), nitrogen dioxide (NO_2), sulfur dioxide (SO_2), ozone (O_3), formaldehyde (HCHO), and radon (Rn) and Rn progeny, as well as various classes of pollutants. The classes of pollutants in which exact composition can vary from one sampling situation to another include fibrous aerosols (of which asbestos is of great concern), biological aerosols, and a number of organic vapors including pesticides and inhalable particulate matter (IP). Table 5 shows the range of measurement techniques, instruments, and methods available for these pollutants.

The remainder of this chapter provides information on pollutant instruments, pollutant methods, air exchange measurements, measurement of environmental parameters, data collection and recording systems, and survey instruments.

TABLE 5. Types of Available Pollutant Measurement Systems by Category

Pollutant	Type	Personal Active	Personal Passive	Portable Active	Portable Passive	Stationary Active	Stationary Passive
Asbestos (fibrous aerosols)	Collector			M*			
	Analyzer			I**			
Biological aerosols	Collector					I	
	Analyzer						
Carbon monoxide	Collector						
	Analyzer	I	I	I		I	
Formaldehyde	Collector	M	I, M	M			
	Analyzer					I	
Inhalable particulate matter	Collector	I, M		M		I	
	Analyzer		I	I		I	
Inorganic/organic particulate contaminants	Collector	M		M		M	
	Analyzer						
Nitrogen dioxide	Collector		I, M				
	Analyzer		I	I		I	
Ozone	Collector						
	Analyzer			I		I	
Pesticides and semivolatile organics	Collector	M		M		M	
	Analyzer						
Radon and radon progeny	Collector						I, M
	Analyzer	I		I		I	I
Sulfur dioxide	Collector						
	Analyzer		I	I		I	
Volatile organics	Collector	M		M		M	
	Analyzer						

*M--One or more methods for this pollutant and measurement category are summarized in Appendix B.

**I--One or more commercially available instruments for this pollutant and measurement category are summarized in Appendix A.

POLLUTANT INSTRUMENTS

Many instruments have been developed for use in the stationary mode as ambient air quality monitors. They can be adapted for indoor air quality monitoring at fixed locations, but they lack mobility and are expensive. Many instruments have been developed for monitoring in workplace environments; they are often personal or portable, but may lack the sensitivity required for monitoring low pollutant concentrations of interest in indoor environments.

Table 6 describes operating principles and characteristics of commercially available instruments for monitoring each pollutant. Detailed descriptions of many pollutant monitoring instruments are included in Appendix A. Instruments that appear on the U.S. Environmental Protection Agency (EPA) List of Reference and Equivalent Methods for these pollutants are also listed in Appendix A. In some cases, portable analyzers based on reference and equivalent methods are available.

For CO, NO_2, SO_2, and O_3, stationary analyzers have been developed to support the monitoring required by EPA's National Ambient Air Quality Standards. Recent advances in electrochemical oxidation cells and supporting electronics have produced personal and portable analyzers for CO, NO_2, and SO_2. Signal integrators and data loggers that can be used to record data from personal analyzers are available. Although such devices have been extensively used with CO personal monitors, they can be used with any device that features a continuous analog output. The signal integrators and data loggers are discussed separately in this chapter. Passive instruments in the form of collectors or analyzers are available for all gas-phase pollutants, with the exception of passive collectors for CO.

Commercially available devices for monitoring HCHO include an automated wet-chemical analyzer and passive collectors. For fibrous aerosols, especially asbestos, users can determine concentrations through sample collection with a portable analyzer. Portable IP analyzers are based on optical scattering and on piezoelectric resonance. Some optical-scattering analyzers are sufficiently miniaturized for personal monitoring of IP. Stationary collectors are also available.

For Rn and Rn progeny, a variety of sophisticated monitors are available. Many are small enough to be considered portable, but the measurement techniques are most often geared to stationary sampling. Two types of passive Rn collectors are available--the TRACK ETCH™ radon detector and two thermoluminescent dosimeters.

Detailed discussion of the many commercially available instruments is beyond the scope of this book. However, because of their importance in indoor air quality monitoring programs, a number of instruments are described in Appendix A to assist the reader in evaluating instruments for particular monitoring needs.

POLLUTANT METHODS

Though a relatively wide range of commercially available instruments for indoor air quality monitoring are identified, the assembly of measurement systems from simple components can also be a practical alternative. In some cases, this approach is the only alternative because suitable instruments are not available for some pollutants. User-configured samplers may also be cost effective if components are reusable and the overall sampling period is such that recycling is feasible.

User-configured methods are also attractive when packaging two or more pollutants into one sampler. The Gage Research Institute Personal Sampler (Mintz et al. 1982),

TABLE 6. Summary of Selected Instruments for Measuring Pollutant Concentrations

Pollutant	Operating principle	Personal, portable, or stationary	Active or passive	Analyzer or collector	Appendix A cross-reference
Asbestos and other fibrous aerosols	Induced oscillation/optical scattering--sample air passes through an oscillating electric field. Fibers are detected by detecting right-angle scattering pulses from larger illumination aligned with the fiber axis.	Portable	Active	Analyzer	A-1
Biological aerosols	Impaction--sample air passes through a series of selective stages (petri dish containing agar); inertial effects cause particles in size ranges of interest to collide with collector surface. Microbial colonies are incubated and counted.	Stationary	Active	Collector	B-1, B-2
CO	Nondispersive infrared (NDIR)--infrared radiation passes through parallel optical cells, one containing sample air, the other containing reference CO-free air. The difference in absorbance relates to CO concentration.	Stationary	Active	Analyzer	EPA Reference Method
	Gas filter correlation (GFC)--infrared radiation passes through a spinning filter wheel that contains a sealed CO reference cell and a nitrogen reference cell. The IR beam then passes through a chamber containing sample air and is detected. The signal difference observed between the nitrogen cell and the CO cell relates to CO concentration.	Stationary	Active	Analyzer	EPA Equivalent Method
	Electrochemical oxidation--sample air passes into an electrochemical cell where oxidation of CO to CO2 produces a signal related to the CO concentration.	Personal Personal Portable	Active Passive Active	Analyzer Analyzer Analyzer	C-3 C-2, C-5 C-1, C-4
HCHO	Wet chemical--HCHO is scrubbed from the sample airstream by a standard reagent solution. Addition of a second reagent forms a distinctive color whose intensity is related to HCHO concentration.	Stationary	Active	Analyzer	F-2
	Sorption/spectrophotometry--HCHO is adsorbed and subsequently quantitated in the laboratory.	Personal	Passive	Collector	F-1, F-3, F-4

(Continued)

TABLE 6. Summary of Selected Instruments for Measuring Pollutant Concentrations (Continued)

Pollutant	Operating principle	Personal, portable, or stationary	Active or passive	Analyzer or collector	Appendix A cross-reference
IP	Optical scattering--sample air passes through a size-selective inlet prior to entering an optical cell. Forward light scattering from controlled light source relates to IP concentration.	Personal Portable	Passive Active	Analyzer Analyzer	I-1, I-3 I-2
	Filtration--sample air passes through a size-selective inlet. Particles in size range(s) of interest are retained on filter(s) for mass determination in the laboratory.	Stationary	Active	Collector	I-4, I-5
	Impaction--sample air passes through a series of selective stages; inertial effects cause particles in certain size range of interest to impact on collector surfaces.	Personal	Active	Collector	I-6
	Piezoelectric resonance--sample air passes through a size-selective inlet. Particles within the size range of interest are electrostatically precipitated onto a quartz crystal. Alterations in oscillation frequency relate to collected mass.	Portable Stationary	Active Active	Analyzer Analyzer	I-7 I-7
NO_2	Gas-phase chemiluminescence--photon emission that accompanies reaction of NO with O_3 is monitored to simultaneously quantify NO and NO_x. NO_x is quantified by first reducing all oxides of nitrogen to nitric oxide (NO). NO_2 is the algebraic difference between NO_x and NO.	Portable Stationary	Active Active	Analyzer Analyzer	N-3 EPA Reference Method
	Triethanolamine (TEA) adsorption--NO_2 is quantitatively sorbed onto treated substrate for subsequent spectrophotometric analysis in the laboratory.	Personal	Passive	Collector	N-1, N-8
	Wet chemical--NO_2 reacts with a reagent system and is quantified colorimetrically.	Personal Stationary	Passive Active	Collector Analyzer	N-4 N-2
	Electrochemical--sample air passes into an electrochemical cell where NO_2-specific reactives produce a signal related to NO_2 concentration.	Personal Portable	Passive Active	Analyzer Analyzer	N-7 N-5, N-6

(Continued)

TABLE 6. Summary of Selected Instruments for Measuring Pollutant Concentrations (Continued)

Pollutant	Operating principle	Personal, portable, or stationary	Active or passive	Analyzer or collector	Appendix A cross-reference
O_3	Gas-phase chemiluminescence--photometric detection of the chemiluminescence resulting from the gas-phase reaction between ethylene and O_3.	Portable Stationary	Active Active	Analyzer Analyzer	O-1 EPA Reference Method
	Gas-solid phase chemiluminescence--photometric detection of the chemiluminescence resulting from the reaction between O_3 and rhodamine-B.	Stationary	Active	Analyzer	EPA Equivalent Method
	Ultraviolet absorption--measurement of the difference in ultraviolet intensity between sample air and reference.	Stationary	Active	Analyzer	EPA Equivalent Method
Rn/ Rn progeny	Filtration/gross alpha counting--Rn progeny collect onto a filter; consequent alpha activity relates to working level (WL).	Personal Portable Stationary	Active Active Active	Analyzer Analyzer Analyzer	R-2 R-4 R-7
	Electrostatic collection/thermoluminescent dosimetry--Rn passes into a special chamber where subsequent progeny (ions) are electrostatically focused onto a thermo-luminescent dosimeter (TLD) chip. Subsequent alpha disintegrations create metastable defects in the TLD, which are deactivated and quantified in the laboratory.	Stationary	Passive	Collector	R-1, R-8
	Alpha scintillation--Rn progeny collect on a filter; Rn is collected in a scintillation flask. Subsequent alpha activity relates to WL (filter sample) and to Rn concentration (scintillation flask).	Portable Stationary	Active Active	Analyzer Analyzer	R-5 (Rn plus WL) R-3 (Rn only)

(Continued)

41

TABLE 6. Summary of Selected Instruments for Measuring Pollutant Concentrations (Concluded)

Pollutant	Operating principle	Personal, portable, or stationary	Active or passive	Analyzer or collector	Appendix A cross-reference
Rn/ Rn progeny (Continued)	Filtration/alpha spectroscopy coupled to electrostatic collection/alpha spectroscopy--Rn progeny (ions) are collected on a filter; subsequent alpha decay relates to WL. Rn passes into a special chamber where subsequent decay ions are electrostatically focused onto a detector; alpha decay relates to Rn concentration.	Stationary	Active	Analyzer	R-6
	Filtration/alpha and beta spectroscopy-- Rn progeny are collected on a filter; subsequent alpha and beta activity of the collected sample relate to WL.	Stationary	Active	Analyzer	R-9
	TRACK ETCH™--alpha-sensitive film registers damage tracks when chemically etched; average Rn concentration is related to the number of damage tracks per unit film area.	Stationary	Passive	Collector	R-10
SO_2	Flame photometric detection (FPD)-- measurement of sulfur-specific emissions from hydrogen-rich air flame.	Stationary	Active	Analyzer	EPA Equivalent Method
	Pulsed fluorescence--measurement of the intensity of the ultraviolet fluorescence of SO_2 excited by a high-intensity light source.	Stationary	Active	Analyzer	EPA Equivalent Method
	Wet chemical--SO_2 reacts with a reagent system and is quantified conducto-metrically or colorimetrically.	Stationary	Active	Analyzer	S-1, EPA Equivalent Method
	Electrochemical--sample air passes into an electrochemical cell where reactions specific to SO_2 produce a signal proportional to concentration.	Personal Portable	Passive Active	Analyzer Analyzer	S-3 S-2

which allows integrated personal sampling of SO_2, NO_2, and particulate matter for up to 10 h, is an example of such an approach. The sampler is built from commercially available components and uses previously tested procedures. This system offers size, portability, and multipollutant sampling that cannot be achieved from any single commercial source.

Methods summarized here represent either standard practices endorsed by a professional organization or widely accepted techniques found in refereed journals. As outlined in Table 7, user-configured methods are available for asbestos, HCHO, IP (including chemical characterization of IP), NO_2, organic pollutants, and Rn. All of these methods involve a collector and laboratory analysis and are summarized in Appendix B. Decisions to pursue any of these approaches should be made only after careful review of supporting literature.

AIR EXCHANGE MEASUREMENTS

The continual transfer of air across the building envelope is an important determinant of indoor pollutant levels. Depending on the prevailing indoor and outdoor concentrations, the outdoor air can either help reduce indoor concentrations or increase indoor concentrations (Chapter 3). When more than one pollutant is involved, both situations may apply at the same time.

The air exchange process results from a complex interplay of many factors (Chapter 3). Some of these factors, such as construction and wind barriers, are essentially constant for a particular building. Other factors, such as wind velocity, indoor-outdoor temperature differences, and mechanical ventilation, are time-varying.

There are two fundamental approaches to measuring air exchange: (1) pressurization techniques, which use measured pressure-flow relationships to evaluate building tightness, and (2) tracer-gas techniques, which use measured concentrations of specially released tracers to evaluate air exchange. Tracer-gas techniques can also be adapted to the problem of measuring airflows from area to area within a building.

Pressurization Techniques

Under normal conditions, the air exchange process is driven by indoor-outdoor pressure differences that amount to a few pascals (Pa). Pressurization techniques involve artificially increasing this pressure difference to evaluate the leakage characteristics of the building envelope. As summarized in Table 8, there are two pressurization methods--fan pressurization and AC pressurization.

Fan Pressurization. The fan-pressurization method has been designated as standard practice by the American Society for Testing and Materials (ASTM 1981a). In this method, leakage characteristics of a structure are measured under controlled pressurization and depressurization. A range of positive and negative indoor-outdoor pressure differences is produced by using a variable-speed reversible fan, which is temporarily installed in an entry doorway. The fan can move large volumes of air into or out of the structure. At a constant indoor-outdoor pressure difference, all air flowing through the fan is compensated by equal flow through available openings in the building envelope. When all controllable external openings such as windows and doors are closed, the resulting data can be used to evaluate the leakage characteristics of the building envelope and thus form the basis for comparisons of relative tightness. This method does not measure infiltration rates directly; rather, it measures the amount of leakage under standard but exaggerated indoor-outdoor pressure differences.

TABLE 7. Summary of Selected Collector-Based Methods for Measuring Pollutant Concentrations

Pollutant	Operating principle	Personal, portable, or stationary	Active or passive
Asbestos and other fibrous aerosols	Filtration--a laboratory analyzes the filters.	Personal	Active
HCHO	Wet chemical--HCHO is scrubbed from the sample airstream by a standard reagent solution. Addition of a reagent forms a distinctive color whose intensity is related to HCHO concentration.	Personal Portable	Active Active
	Sorption/spectrophotometry--HCHO is adsorbed onto substrate and subsequently desorbed and quantitated in the laboratory.	Personal Portable	Passive Active
IP	Filtration--sample air passes through a size-selective inlet. Particles in size range(s) of interest are retained on filter(s) for mass determination in the laboratory.	Personal Portable	Active Active
Metals and other inorganic particulate constituents	Filter collection/laboratory analysis--inorganic constituents are collected on a suitable filter. Metals may be quantitated by atomic absorption spectroscopy, neutron activation analysis, or proton-induced X-ray fluorescence. Nitrates and sulfates can be determined spectrophotometrically.	Personal Portable Stationary	Active Active Active
NO_2	Triethanolamine (TEA) adsorption--NO_2 is quantitatively sorbed onto treated substrate for subsequent quantitation in the laboratory.	Personal	Passive
Pesticides and other semivolatile organics	Sorbent collection/laboratory analysis--semivolatile organics are collected by passing sample air through polyurethane foam. In the laboratory, compounds are extracted for chromatographic quantitation.	Personal Portable Stationary	Active Active Active
Polyaromatic hydrocarbons and other organic particulate constituents	Filter collection/laboratory analysis--particle-associated organics are collected on a suitable filter. Organic constituent may be quantified through a number of techniques.	Personal Portable Stationary	Active Active Active
Rn/ Rn progeny	Sorption/gamma activity--Rn is adsorbed onto activated charcoal; subsequent gamma activity is related to average Rn concentration.	Stationary	Passive
Volatile organics	Sorbent collection/laboratory analysis--volatile organics are collected by passing sample air through a suitable adsorbent column. In the laboratory, compounds of interest are desorbed for chromatographic quantitation.	Personal Portable Stationary	Active Active Active

TABLE 8. Summary of Pressurization Techniques for Measuring Building Tightness

Operating principles	Key relationships*
Fan Pressurization--a constant pressure difference is created across the building envelope using a fan. Pressure-flow relationships for positive and negative pressure differences are evaluated to determine building tightness.	$Q = K\ \Delta P^n$ $L = K\ \sqrt{\dfrac{P}{2\Delta P}}$
AC Pressurization--an oscillating pressure difference is created across the building envelope using a piston to vary the indoor volume. Relationships between piston cycling and pressure cycling in the indoor airspace are evaluated to determine building tightness.	$L = -\sqrt{\dfrac{\rho}{2\Delta P_r}}\ \ \dfrac{\left< Qd\ \dfrac{\Delta P}{\Delta P_r} \right>}{\left< \left(\dfrac{\Delta P}{\Delta P_r}\right)^{n+1} \right>}$

*See text for nomenclature.

ASTM-specified equipment includes the following major components:

- Air-moving equipment--capable of sustained flows up to 5 100 m^3/h (3 000 ft^3/min at a constant rate

- Pressure-measuring device--capable of measuring pressure differentials with an accuracy of ±2.5 Pa (±0.01 in of water)

- Airflow-measuring system--to measure flows to within ±6 percent over the operating range of the air mover

- Airflow-regulating system--to regulate and maintain flows induced by the air-moving equipment to within 20 percent or less.

These components may be integrated to form a blower-door assembly to facilitate mounting the unit in the doorway and to offer convenient placement of readouts. Some commercial sources for obtaining complete blower-door assemblies that meet ASTM requirements are as follows:

GADZCO, Inc.	HARMAX Corporation	Retrotec Energy Innovations, Ltd.
209 Vetterlein Avenue	6224 Orange Street	176 Bronson Avenue
Trenton, NJ 08619	Los Angeles, CA 90048	Ottawa, Ontario, Canada K1R6H4
(609) 586-6747	(213) 936-2673	(613) 234-3368

The desired range of induced pressure differences is from 12.5 Pa to 75 Pa (0.05 to 0.3 in/water), in increments of 12.5 Pa. In some cases, leakage rate may exceed fan capacity; nonetheless, a minimum of five data points on each side of zero is desirable. Each data point consists of the measured pressure difference (Pa) and the corresponding fan flow (m^3/h). In addition to measurements of flow rate and pressure differentials, measurements of onsite windspeed and indoor-outdoor temperatures are useful. Preferred environmental conditions include winds of 2.5 m/s (5 mi/h) or less and indoor-outdoor temperature contrasts of 11 °C or less to minimize the environmentally induced pressure differential. Winds in excess of 4.5 m/s (10 mi/h) should be avoided, and data collection under winds between 2.5 and 4.5 m/s should be approached with caution because gusting can distort pressure-to-flow relationships. In cold weather applications, extreme caution must be exercised to avoid bringing in cold outdoor air.

All measured flows are converted to standard conditions (101.3 kPa pressure, 21.2 °C temperature, 1.2 kg/m^3 air density) and plotted against corresponding pressure differences. ASTM estimates the uncertainty in the measurements at a given pressure difference at 10 percent or less.

The fan-pressurization data can be used to determine the effective leakage area, which can be used as an index value for comparison purposes. Using the approach developed by Sherman and Grimsrud (1980), the pressure differential and flow relationships are assumed to be similar to those for a perfect orifice; data are fitted to the general equation:

$$Q = K \ (\Delta P)^n \qquad\qquad\qquad (1)$$

where Q is the flow (m^3/h) and ΔP is the pressure difference (Pa). The constants K and n are determined empirically to gain a best fit of the data. Separate regressions are usually performed for positive and negative pressurization to allow for recognition of asymmetric leakage. This equation can then be used to calculate the flow at any convenient pressure difference. The effective leakage area, L (m^2), can then be calculated from the expression:

$$L = Q \ \sqrt{\frac{\rho}{2\Delta P}} \qquad\qquad\qquad (2)$$

where ρ is the density of air (1.2 kg/m^3). The effective leakage area should not be confused with the air exchange rate; rather, it is an estimate of the aggregate size of the openings through which infiltration may occur at rates determined by a variety of influences.

Effective leakage area can be determined for any measured or interpolated pressure difference. Sherman and Grimsrud (1980), for example, use the interpolated flow at 4 Pa. Recently, it has become a common practice to use the measured air leakage rate at 50 Pa pressure difference as a reference point for comparisons of air tightness in buildings (Liddament and Thompson 1983). Regardless of the index pressure difference used, any data report should include all measured values.

An additional benefit of the fan-pressurization method is the opportunity to identify individual routes of leakage during any tightening procedure. Identification is accomplished by either pressurizing or depressurizing a structure with the blower door and using a smoke stick or other visual tracer to note the flow through leakage sites. Similarly, infrared thermography of parts of the building combined with use of a blower door can assist in precisely identifying leakage sites.

AC Pressurization. AC pressurization creates an oscillating pressure difference across the building envelope using a piston or bellows drive to alter the indoor volume at a known frequency. The amplitude and phase of the indoor pressure with respect to the volume drive are related to the tightness of the building envelope. By operating in regimes of pressure difference that are of the same magnitude as the weather-induced pressure differences that drive natural air exchange, application of the technique circumvents the need for the large airflows that accompany fan pressurization tests.

The technique involves a volume drive to produce a sinusoidal change in indoor volume at an appropriate amplitude (0.01 to 0.05 m^3) over a useful frequency range (0.1 to 4 Hz). The pressure oscillations produced by the volume drive are monitored with a low-frequency microphone.

Developmental work at Syracuse University (Graham 1977; Card et al. 1980) encountered low accuracy as compared to fan pressurization. Independent work at the Lawrence Berkeley Laboratory (Grimsrud et al. 1980a) applied phase-sensitive synchronous detection techniques that extract a component of the pressure signal that is in phase with the oscillating source to increase accuracy and simplify data interpretation. Modera and Sherman (1985) report encouraging results from extensive field tests of a prototype AC pressurization device employing synchronous detection. Although further testing and development is required before AC pressurization systems can be manufactured commercially, the technique offers a viable alternative to fan pressurization.

Expression of leakage area from AC pressurization is similar to that from fan pressurization (equation 2); because the pressure-flow relationships oscillate, cycle averages are used to characterize the system. At a particular reference pressure difference (ΔPr), the effective leakage area is related to the time rate of change of the drive displacement (Qd, m^3/s) by:

$$L = -\sqrt{\frac{\rho}{2\Delta P_r}} \quad \frac{\left< Qd\, \frac{\Delta P}{\Delta P_r} \right>}{\left< \left(\frac{\Delta P}{\Delta P_r}\right)^{n+1} \right>} \tag{3}$$

where $<\ldots\ldots>$ denotes a cycle average.

Because the technique cannot determine the flow exponent directly, leakage area measurements remain sensitive to estimates for n. From a large data set of fan pressurization data, Modera and Sherman found a mean value of 0.65 for the flow exponent with a standard deviation of 0.09.

Substituting this empirical value gave leakage area results that were in good agreement with fan pressurization tests conducted on the same houses. AC pressurization results were, however, consistently lower on the average (by approximately 14 percent) than fan pressurization tests. Additional experiments with fireplace dampers and window openings showed that AC pressurization treats large leaks greater than some critical size as if they were of that critical size.

These tests did confirm that the drive volume selected (0.05 m^3) was able to provide sufficient pressures without resorting to higher frequencies that are encumbered by resonance effects. It was also found that the location of the microphone did not affect the measured pressure signal--even on the second story of two-story houses.

Tracer-Gas Techniques

All tracer-gas techniques are fundamentally tied to the mass balance considerations introduced in Chapter 3. Ideally, a tracer is inert and not normally present in the indoor or outdoor atmosphere. The indoor concentration of a tracer released into indoor airspace and well-mixed can be described by the following equation:

$$\frac{dC_{in}}{dt} = \frac{S}{V} - \nu C_{in}. \tag{4}$$

By knowing or controlling key aspects of the mass balance, the air exchange rate can be measured. As shown in Table 9, three basic techniques are in general use: (1) concentration decay, (2) constant injection, and (3) constant concentration. Regardless of which type of tracer-gas technique is selected, additional measurements of winds, indoor and outdoor temperature and humidity, and barometric pressure are recommended. Because air exchange rates attributable to infiltration can vary substantially due to changing environmental conditions, tracer-gas tests should be performed under a variety of conditions, if a measure of average infiltration is desired.

TABLE 9. Summary of Tracer–Gas Techniques for Measuring Air Exchange

Operating principles	Key relationships*
Concentration Decay--a small amount of tracer gas is introduced and mixed into the indoor airspace. The resulting decay of tracer gas concentration indoors, due to entry of tracer-free outdoor air, is related to the air exchange rate.	$C_{in,t} = C_{in,t_i}\, e^{-\nu(t-t_i)}$
Constant Injection--tracer gas is released to the indoor airspace at a constant rate. Changes in indoor concentration are related to tracer release rate and air exchange rate.	$\bar{\nu} = \dfrac{\bar{S}}{VC_{in}}$
Constant Concentration--tracer-gas release is controlled to maintain a constant concentration. Air exchange rate is related to tracer release rate.	$C_{in} = \dfrac{S}{\nu V}$

*See text for nomenclature.

In general, an appropriate tracer gas has the following characteristics:

● It is easily and inexpensively measured at low concentrations and over short sampling times.

● It is not a normal constituent of air or it normally persists at concentrations many orders of magnitude below concentrations for air infiltration measurements.

● Its measurement technique is interference free with regard to normal atmospheric constituents and thermodynamic conditions.

● It is inert, nonpolar, and not absorbed.

● It presents no safety or health hazard to occupants or operators.

As summarized in ASTM E741, no single tracer gas satisfies all of these conditions. However, as long as precautions are taken to ensure that initial concentrations are acceptably low, a number of gases are acceptable. Recommended practice is to restrain maximum concentrations to at least a factor of 4 below accepted limits. Under no circumstances should initial tracer-gas concentrations exceed the Occupational Safety and Health Administration (OSHA) time-weighted average for substances included in the latest OSHA-controlled list. Sulfur hexafluoride (SF_6) is one of the most commonly selected tracer gases; other tracer gases include nitrous oxide (N_2O), carbon dioxide (CO_2), and ethane (C_2H_6). Use of these and other tracer gases is discussed in Grimsrud et al. (1980), Harrje et al. (1982), and Lagus and Persily (1985).

Concentration decay: The concentration decay technique (also called tracer-gas dilution) has been designated as a standard practice by the American Society for Testing and Materials (ASTM 1981b). In this technique, a small amount of tracer gas is injected into the indoor airspace and thoroughly mixed. Because there are no further releases of tracer gas (S = 0), the mass balance can be simplified to:

$$\frac{dC_{in}}{dt} = - \nu C_{in} \tag{5}$$

with the solution,

$$C_{in,t} = C_{in,t_i} \, e^{-\nu \Delta t} . \tag{6}$$

Thus, indoor concentrations "decay" with time as the exfiltrating air removes the tracer. The air exchange rate can be calculated directly by rearranging this equation to form:

$$\nu = \frac{1}{\Delta t} \ln\left(\frac{C_{in,t_i}}{C_{in,t}}\right). \tag{7}$$

When a succession of data points is obtained, the air exchange rate can be estimated graphically from a log-linear plot of concentration versus time or calculated through log-linear regression or finite difference methods to achieve a best fit.

The general procedure involves releasing tracer gas at one or more points in sufficient quantities to produce useful initial concentrations. The method of release and quantities involved depend on the internal volume of the structure, the configuration of the air-handling system, estimates of allowable versus useful concentrations, and sensitivity of the detection system. In buildings that have central air-handling systems, releases may be introduced directly to the intake. Otherwise, releases can be made from multiple points and mixed with portable fans. In general, at least 30 min should be allowed for mixing prior to formal sampling.

Tracer-gas samples should be taken periodically from two or more widely spaced locations on each story. Sampling methods range from simple grab sampling with later analysis in the laboratory to sophisticated real-time analyzers that produce a time-series of tracer gas concentrations. Sampling methods are outlined in Liddament and Thompson (1983), in Harrje et al. (1982), and in Lagus and Persily (1985).

Harrje et al. (1982) have developed and tested a very simple technique using flexible plastic bottles. In this approach, one set of bottles is used to release the tracer gas, and after mixing is complete, a different set of bottles is used to acquire a set of samples during decay. Tracer gas is released by first loosening, but not fully removing, the cap of a bottle previously loaded with tracer and gently squeezing the bottle while moving throughout the structure. After mixing is complete, a number of samples are acquired during the dilution period using clean bottles. Samples are then returned to the laboratory for analysis. The caps of the sample bottles feature a septum to allow drawing one or more aliquots for direct injection into the gas chromatograph (GC). The squeeze-bottle method has been tested in residential settings where residents have performed the tracer release and sampling. The method has not been tested in larger buildings.

Syringes are frequently used to acquire grab samples. With SF_6 tracers, in particular, samples can be injected directly into a GC for analysis. Although sampling syringes may be considered reusable, care should be exercised in purging SF_6 from syringes previously used; purging should be verified analytically.

Syringe sampling can also be automated. One manufacturer* offers a portable, battery-powered sampling device that automatically collects up to 12 samples on a user-defined time base and accepts syringes up to 1 200 mL in volume. Use of such syringe samplers provides a practical method for measurement of air exchange rates over a short (3 to 6 h) duration (Nagda et al. 1986).

Constant injection. Concentration decay tests usually run out of tracer in a few hours. For more extended testing, additional injection and mixing cycles can generate more concentration decay periods. An alternative is to constantly release tracer and monitor indoor tracer-gas concentrations. Air exchange rates may be evaluated from the mass balance described in equation 4. The general solution for a well-mixed system is:

$$C_{in,t} = C(t_i)e^{-\nu\Delta t} + \frac{\overline{S}}{\nu V}(1 - e^{-\nu\Delta t}). \tag{8}$$

Over suitably long averaging periods, equation 8 reduces to:

$$\overline{\nu} = \frac{\overline{S}}{VC_{in}}. \tag{9}$$

Rather elaborate systems such as that developed and tested by Condon et al. (1978) have been used to implement equation 8. A truly practical version of the constant injection technique based on equation 9 has been developed and tested at the Brookhaven National Laboratory using diffusion-based release and sampling of perfluorocarbon tracers (PFTs) (Dietz and Cote 1982a, 1982b). In this method, a bullet-sized container releases the PFT at a typical rate of 10 nL/min. Sampling is carried out using a 4-mm diameter capillary adsorption tube (CAT) that collects the tracer by diffusion. CAT samplers are activated by simply removing the end cap; replacing the cap halts sampling. Average concentrations are determined in the laboratory by gas chromatography. Source release rates are verified through reweighings.

---

*Demarry Scientific Instruments, Ltd., SE 1122 Latah Street, Pullman, Washington 99163, (509) 332-8577.

One PFT source should be used for approximately every 50 m^2 (500 ft^2) of living space; attention should be paid to the floor plan to recognize a need for added sources. Sources should be near outside walls to take advantage of mixing patterns. Samplers should be located at least 1.5 m (5 ft) from any PFT source. Sampling may proceed for periods as short as a day, or may be extended over a number of weeks, if necessary; the upper limit has not been firmly established. Recommended mixing time prior to sampling is 8 h. This delay could present a problem in logistics because the procedure would require a return visit or the involvement of a resident to initiate sampling at the proper time. However, in situations where this delay period is very small compared to the total sampling period (i.e., sample period ≥ 1 week), sources and samplers can be activated simultaneously without significantly affecting data. Special precautions must be taken to avoid contamination of samples; CAT samplers and PFT sources should be isolated from each other during storage and shipping.

Constant concentration. In the constant concentration approach, automated equipment is required to continually analyze tracer-gas concentrations and, based on losses due to air exchange, inject additional tracer to maintain a constant concentration. Because the time rate of change for the tracer is held to zero, equation 4 is simplified to:

$$\nu = \frac{S}{VC_{in}} . \qquad\qquad (10)$$

The constant concentration technique offers an advantage over other single-tracer techniques in that air exchange rates can be measured simultaneously for individual zones within a building. As reviewed in Harrje et al. (1985) and in Lagus and Persily (1985), the constant concentration method has been successfully used in a variety of buildings with as many as 10 zones.

Although this technique is useful for evaluating air exchange rates with outdoors on a zone-specific basis, the flows among zones that are important to indoor air quality cannot be resolved. Furthermore, automating the constant-concentration technique requires rapid tracer analysis and feedback for compensatory injection. Small fans are often incorporated to promote mixing within the zone; however, if fan-assisted measurements are to coincide with pollutant monitoring, the data may not fully reflect normal conditions for some pollutants.

Multizone tracer-gas methods. The tracer-gas methods discussed above apply to situations that do not require a quantitative depiction of air motions within the building. However, for some indoor air quality problems, treating the building as a single, well-mixed chamber can misrepresent conditions. For example, Rn concentrations in rooms on the main floor are determined by the airflows among the basement, main floor, and outdoors.

The tracer-gas dilution method can be adapted to measure internal airflows by injecting tracer into one zone and monitoring tracer concentrations in both zones (Hernandez and Ring 1982). Tracer concentrations in the injection zone fall off rapidly at first as the tracer moves to the second zone as well as to the outdoors. Similarly, tracer concentrations rapidly increase at first in the second zone as tracer moves in from the injected zone. This transient period is followed by a so-called dominant period where tracer concentrations in both zones decline. Airflows between indoor zones and between each zone and the outdoors are calculated from simultaneous mass balance equations.

Single-tracer injection/dilution methods for dual-zone analysis can be limited by a number of factors. For example, Harrje et al. (1985) cite instrumentation as a strong potential limitation. When the two zones are poorly coupled, tracer concentrations are much lower in the second zone; it may not be possible for concentrations in both zones to remain within the analytical range sufficiently far into the dominant period. The same problem can arise when the zones are strongly coupled, but one is much leakier than the other. Another weakness of this approach is that the mathematics assumes that conditions remain constant throughout the entire period (transient plus dominant). Thus, the validity of the results may be questionable if many hours are required for the test.

Multiple-tracer methods represent perhaps the most straightforward approach to examining airflows within buildings. A different tracer is released in each zone; all tracers are then monitored simultaneously in all zones. Because each tracer is identified with a particular zone, airflows among the indoor zones can be determined through mass balance equations.

Multiple tracers can be used in the injection and decay mode as well as in the constant-injection mode. Complete mathematical developments of mass balance relationships that apply to multiple tracers are described by Sinden (1978) and Sandberg (1984).

The diffusion-based constant injection system using PFTs has been successfully field tested in the multiple-tracer mode in single-family residences, apartment buildings, and large commercial buildings (Dietz et al. 1985). Four PFT gases are available as tracers. Most single-family residences can be evaluated using no more than three zones (basement or crawl space, main floor, and second floor). Similarly, airflows in even complex buildings can be measured using three or four tracers with an assumption that direct flows between nonadjacent zones are negligible.

MEASUREMENTS OF ENVIRONMENTAL PARAMETERS

The important environmental parameters in indoor air quality monitoring include windspeed, wind direction, air temperature, humidity, solar radiation, and barometric pressure. Table 10 highlights the operating principles of common environmental transducers. Typical performance characteristics of these transducers are listed in Table 10. The consolidated catalogs and resource directories listed in Chapter 7 can be used to identify manufacturers of transducers for measurement of environmental parameters. References listed in Chapter 7 provide useful insights regarding operating requirements, advantages, and limitations in practical use.

DATA COLLECTION AND RECORDING SYSTEMS

Data handling is an important aspect of the monitoring effort. The advent of the microprocessor has produced exciting capabilities in the collection, retrieval, and processing of data (Conway 1984; Cooney 1985). Two broad categories of data-recording equipment are discussed here: (1) data loggers--devices that read and store one or more channels for later playback or retrieval and (2) general purpose microprocessors that may be easily modified to allow storage and later playback or retrieval.

As shown in Table 11, data loggers range from small, single-channel integrating dosimeters that must be read manually to hand-calculator-sized devices that feature sophisticated programmability. These compact data loggers meet general requirements for personal/portable monitoring. For stationary applications, data loggers capable of treating hundreds of channels are available.

TABLE 10. Summary of Common Transducers for Environmental Measurement

Environmental parameter	Operating principle	Typical performance characteristics				Comments
		Output	Accuracy	Resolution	Range	
Windspeed	Voltage conversion--using an electric generator, the rotary motion of cups or propellers is converted to a voltage that corresponds to windspeed.	Analog voltage	±5%	0.5 m/s	0 to 50 m/s	dc generator most often used; ac generators are limited at lower windspeeds.
	Frequency conversion--using a slotted window to intercept a light beam, the rotary motion of cups or propellers is converted to a frequency that corresponds to windspeed. The frequency is then converted to a corresponding analog voltage.	Analog voltage	±5%	0.5 m/s	0 to 50 m/s	Because the principal signal is frequency dependent, separation distances between transducer and signal conditioner can be large without suffering transmission losses.
Wind direction	Resistance conversion--directional changes of a vane are registered through a single or double potentiometer whose rotary position (and thus resistance) is aligned with direction.	Analog voltage	±2°	1°	0 to 360° (0 to 540°)	Dual potentiometer configurations allow a 0-540° output range to avoid sharp signal changes under northerly winds.
Air temperature	Thermistor--the resistance of a thermally sensitive semiconductor varies inversely with absolute temperature. A composite sensor of two or more thermistors along with fixed resistors provide linear response.	Analog voltage	±0.2 °C	0.1 °C	-50 to 50 °C	With proper sensor configurations, temperature gradients can be directly measured with an accuracy of 0.1 °C.
	Resistance temperature detector (RTD)--the electrical resistance of a pure metal (platinum is mostly used) increases with temperature.	Analog voltage	±0.3 °C	0.1 °C	-200 to >50 °C	RTDs operate at a much lower resistance-to-temperature ratio than thermistors and may require special circuitry to compensate for lead resistance errors.
	Thermocouple--a small, temperature-dependent current is generated at the junction of two dissimilar metals. Temperature can be determined by comparing this current to that generated by the same metals at a temperature-controlled reference junction.	Analog voltage	±1 °C	0.1 °C	Very wide range	

(Continued)

53

TABLE 10. Summary of Common Transducers for Environmental Measurement (Continued)

Environmental parameter	Operating principle	Typical performance characteristics				Comments
		Output	Accuracy	Resolution	Range	
Air temperature (continued)	Deformation--thermal contraction/expansion of working material is mechanically translated to pen position on a graph.	Analog chart, analog signal	±1 °C	0.5 °C	-20 to 40 °C	This transducer is often allied to a recording hygrometer to form a hygrothermograph. Output may be mechanically linked to chart pens (analog chart) or extension/distension may be converted electrically (analog signal) using strain gauges or potentiometers.
Humidity/dewpoint	Psychrometry--the dewpoint or relative humidity (RH) is calculated from the temperature reduction due to evaporation. Two ventilated thermometers, one moistened through a wick (wet bulb) and the other maintained at ambient (dry bulb) are required.	Dictated by thermometry	Dictated by thermometry	Dictated by thermometry	All ambient conditions	At 23 °C and 50% RH, thermometry must be accurate to 0.1 °C to gain ±1% accuracy of measurement. Reducing accuracy of thermometer to 0.5 °C reduces measurement accuracy to ±5%. Measurement may be carried out using any type of thermometer that can sustain conditions of measurement. Measurement is limited by freezing conditions without converting to an antifreeze solution (and another reduction equation).
	Deformation hygrometry--dimensional changes of hygroscopic material are related to RH.	Analog chart, analog signal	±5% RH	5% RH	0 to 100% RH	Human hair is the most commonly used material. This transducer is often allied to a recording deformation thermometer to form a hygrothermograph.
	Electrochemical hygrometry--chemical or electrical changes due to sorption of water vapor are related to RH or dewpoint at a measured temperature.	Analog signal	±0.5 °C dewpoint	0.1 °C	-40 to +50 °C	Lithium chloride is the most frequently used sensing material. Colocated thermometry is usually of RTD type but may be any sensor calibrated for proper dewpoint-to-temperature relationships.
	Chilled mirror--a mirror is chilled thermoelectrically until dew (or frost) forms. Formation of dew (or frost) is sensed optically. Mirror temperature is related to dewpoint.	Analog signal	±0.2 to ±0.4 °C	0.1 °C	-50 to +50 °C	Temperature sensing may be carried using linear thermistors or RTDs. Although complex and expensive compared to other systems), this transducer is often considered to be a functional standard.

(Continued)

TABLE 10. Summary of Common Transducers for Environmental Measurement (Concluded)

Environmental parameter	Operating principle	Typical performance characteristics			Comments
		Output	Accuracy	Resolution	Range	

Environmental parameter	Operating principle	Output	Accuracy	Resolution	Range	Comments
Solar radiation	Pyranometer--measures solar radiation received on a horizontal surface from whole atmosphere using a thermopile or a photocell.	Analog signal	±3 to 10%	0.1 to 1.0 mW/cm²	0 to 150 mW/cm²	
	Pyrheliometer--measures intensity of direct solar radiation using a thermopile or a photocell.	Analog signal	±3 to 10%	0.1 to 1.0 mW/cm²		Requires solar-tracking drive to maintain alignment with the sun.
Barometric pressure	Aneroid barometer--measures expansion/contraction of a partially evacuated capsule in response to hanging pressure.	Analog chart, analog signal	±0.2 mbar (0.005 in Hg)	0.25 mbar (0.01 in Hg)	945 to 1 045 mbr (27.9 to 30.1 in Hg)	Output may be mechanically linked to chart pen (analog chart) or expansion/ contraction may be converted electrically (analog signal) using strain gauges or potentiometers.
	Mercury barometer--measures the height of a mercury column that balances column weight against atmospheric pressure.	Manual analog	±0.1 mbar (0.003 in Hg)	0.1 mbar (0.003 in Hg)	675 to 1 050 mbr (20 to 31 in Hg)	Units are not generally portable; adapts slowly to changing temperature.

TABLE 11. Examples of Data Logging Systems

Manufacturer/ Model/Price*	Size (cm) L W H			Weight (kg)	Power	Nominal operating period	No. analog channels	Internal capacity	Playback/ external storage/output
Custom Instrumentation 1027 Euclid Street Santa Monica, CA 90403 (213) 393-4760 Model: Data Logger $300	11	7	5	0.24	9-V battery	40 h	1	--	4-digit LED display (readout only)†
ENDECO, Inc. 13 Atlantis Drive Marion, MA 02738 (617) 748-0366 Model: Type 1069 $2 900	25	8	8	1	8 AA alkaline batteries	up to 1 yr	8	128 K	RS-232
Metrosonics, Inc. P.O. Box 23075 Rochester, NY 14692 (716) 334-7300 Model: dl701 $1 200	18	9	3	0.6	9-V battery	100 h	1	3.6 K	RS-232, LCD display
Metrosonics, Inc. P.O. Box 23075 Rochester, NY 14692 (716) 334-7300 Model: dl721 $2 600	28	17	5	2.3	6-V rechargeable battery	200 h	8	7.2 K	RS-232, LCD display
Campbell Scientific, Inc. P.O. Box 551 Logan, UT 84321 (801) 753-2342 Model: 21X Micrologger $2 300	21	15	8	2.8	8 D batteries	Months	8, 16	up to 23 K	RS-232, LCD display
Monitor Labs, Inc. 10180 Scripps Ranch Blvd. San Diego, CA 92131 (619) 578-5060 Model: Model 9350 $5 000	56	48	18	18.2	115/230 V ac	Unlimited	up to 1 040	16 K	RS-232

*General level of 1985 cost for comparison; does not include options, accessories, or peripherals.
†System stores a running time-weighted average of signal voltage that can be read from display. Built-in digital clock allows manual readings by monitoring subject.

General-purpose calculators and computers can be easily converted for data logging through commercially available accessories and peripherals. In many cases, this hardware investment competes favorably with that for prepackaged data loggers and the devices retain their original capabilities when not collecting data.

Fitz-Simmons and Sauls (1984) describe such an approach for personal monitoring using Hewlett-Packard (HP) equipment. The system consists of an HP-41CV hand-held calculator with various accessories available from HP to increase storage registers and to provide clock functions and interfacing. Programming allows storage of sequential averages or instantaneous values with playback through the display for manual translation or through an RS-232 compatible port for computerized applications.

Desk-top or personal computers can be modified through accessory boards and peripherals to log data. A wide range of digitizers, scanners, and related accessories are available for systems such as the IBM PC and the Apple II series. Product reviews and manufacturer contact information can be obtained from numerous periodicals on personal computers. If monitoring can be conducted in a stationary mode, users can consider large numbers of data channels and records that are directly compatible with spreadsheet programs such as Lotus 1-2-3™ or statistical software packages such as SPSS-PC™.

SURVEY INSTRUMENTS

Development of survey instruments for indoor air quality studies is sometimes given secondary attention in comparison to other design aspects. In such an instance, a survey instrument or instruments may be hastily prepared, almost as an afterthought, using questions developed by other investigators as precedents. The assembly of these instruments deserves a more deliberate and systematic approach (Koontz and Nagda 1985). If properly developed, survey instruments can provide substantial information to explain variations in measured pollutant levels and related parameters such as air exchange.

Two basic types of survey instruments can be distinguished--questionnaires and activity logs. Questionnaires consist of a series of questions or information items that are administered to building occupants either in person, over the telephone, or through the mail (i.e., self-administered). Questionnaires can also be completed by field technicians, without the assistance of occupants, on the basis of observations or physical measurements.

Activity logs are forms completed by study participants to indicate the times when certain activities take place such as turning an appliance on or off, opening a window, or leaving a building. An entry is usually made on the log each time an activity of interest is started or completed, but each log can also be filled out using short-term recall (that is, by entering all activities for one day at the end of that day).

The following four steps are vital to the development and assessment of survey instruments:

● Choosing factors that need to be addressed to satisfy monitoring and study objectives

● Classifying factors to capitalize on different opportunities for obtaining pertinent information

● Determining the appropriate wording of questions to ensure that accurate, consistent, and complete information is obtained

● Judging the relative efficacy of survey instruments or specific questions after data collection has been completed.

Choice of Factors

The mass balance equation for indoor air quality described in Chapter 3 provides a systematic framework for identifying the types of factors that need to be quantified. If all parameters included in the equation could be measured directly, there might be no need for survey instruments; however, this is rarely the case.

Each parameter in the mass balance equation can be related to various groups of factors that can be quantified, either through direct physical measurements or observations or through questions posed to building occupants. As an example, consider one term of the mass balance equation--the air exchange rate. This term is influenced by the infiltration of outdoor air, by natural ventilation practices, and by mechanical ventilation. Thus, structural characteristics, weather, and occupant activities have a role in determining the prevailing air exchange rate at any point in time. Similarly, these and other factors, such as indoor sources or furnishings, affect indoor air quality. A number of specific factors related to each mass balance term is listed in Table 4 in Chapter 3.

The specific factors of importance and the manner in which they influence indoor air quality certainly can vary from pollutant to pollutant; thus, the choice of factors to be quantified is strongly tied to the choice of pollutants to be monitored. If more than one pollutant is to be monitored, then the choice of factors should proceed independently or in parallel for specific pollutants. The specific monitoring objectives and timeframe for pollutant measurements also influence the choice of factors. For example, more detail on occupant activities might be required for a week-long measurement with a continuous analyzer than for a month-long measurement with a sample collector.

Classification and Characterization of Factors

One important type of classification is determining whether the factors are static or dynamic. Static factors are those characteristics that typically do not change over time or that change infrequently such as the structural properties of a building or the types of appliances that it contains. In contrast, dynamic factors, such as occupant habits or practices, typically vary over time. The static and dynamic distinction is important in the development of survey instruments because these two classes of factors require different methods for obtaining information. In the case of factors affecting air exchange, one static factor is the age of the building, which can be directly characterized by asking when it was built. One dynamic factor is the frequency of opening windows, which could be characterized as the usual frequency during a specific season or the actual frequency during a recent week, based on the occupant's recall. A real-time estimate of this frequency could also be obtained, for example, by asking the occupant to record the times when each window was opened or closed over a particular week.

In any monitoring effort, opportunities to characterize factors typically arise at three stages:

● Premonitoring--the period during which participants of a monitoring study are solicited and enrolled

● Monitoring--the period during which air quality measurements are taken

● Postmonitoring--the period that immediately follows the completion of air quality measurements.

Defining the three stages is important because it affects (1) whether characterization of static factors requires occupant assistance or only the observations
of a field technician, (2) whether general practices or specific details need to
be characterized for dynamic factors, and (3) whether information concerning
practices can be determined concurrently with monitoring or retrospectively.

The nature of factors and stages of monitoring provide a matrix for obtaining
information (Table 12). A number of important points should be recognized:

● Some characterization of static factors can be obtained from occupants in
  advance of monitoring through personal or telephone interviews or by self-
  administered questionnaires, but occupants can be relieved of some of this
  burden if technicians observe certain characteristics and obtain supplemental
  information from occupants either during or after the monitoring.

● Occupants can provide a limited description of general practices in advance of
  monitoring, but actual practices that can influence monitoring results need to
  be characterized on a real-time or short-term-recall basis; this information
  is generally recorded by occupants using activity logs; some of the occupant's
  burden for this activity can be relieved if technicians periodically pose
  supplemental questions during or after monitoring.

● After monitoring has been completed, it is also possible to obtain clarifying
  or supplemental information from occupants.

**TABLE 12. Methods of Characterizing Factors, Classified According to Nature
of Factor and Stage of Monitoring**

Stage of monitoring for characterizing factors	Nature of factor requiring characterization	
	Static	Dynamic
Premonitoring	Description by partici-pants of general characteristics of structure and contents	Description by partici-pants of general habits or practices
Monitoring	Observations by tech-nicians or questions from technicians to participants concerning more detailed character-istics of structure and contents	Recording by participants of actual practices on a real-time or short-term-recall basis
Postmonitoring	Clarification from participants concerning previously obtained information or provision of limited supplemental information	Supplemental information concerning participants' real-time records asked by field technicians on a short-term-recall basis

Generally, it would be easier to obtain information on static factors than on
dynamic factors at the premonitoring stage. On the other hand, supplemental
information obtained at the postmonitoring stage would usually be more valuable
for dynamic factors than for static factors.

Formulation of Questions

Questions to characterize factors affecting indoor air quality need to be formulated in such a way that the information sought is clear to all potential respondents, regardless of age, education level, occupation, or socioeconomic status.
Moreover, the specific formulation can vary depending on the mode by which questions are administered--mail, telephone, or in person. The telephone and personal
routes certainly provide a greater opportunity to interact with respondents and
to clarify or probe their responses.

Principles that underlie the formulation of questions and the design of questionnaires are a major component of the work of professionals engaged in survey
research. Important considerations include determining the order in which
questions are to be asked; choosing the degree of precision sought by a question
and deciding whether closed-ended or open-ended responses are best suited for
obtaining the needed precision; ensuring that response categories, when used, are
mutually exclusive and collectively exhaustive; constructing clear skip patterns
(that is, paths to subsequent questions that are dependent on the response to a
specific question); and asking questions in such a way that responses will not be
biased.

Some guidelines for formulating mail and telephone questionnaires are provided
in a publication by Dillman (1978); his examples are mainly in the context of
sociological research but the principles have much broader application. Some
useful precedents for the wording of questions concerning building characteristics are contained in questionnaires used by Federal agencies, such as the
1980 Census of Population and Housing and the Annual Housing Surveys conducted
by the Bureau of the Census and the Residential Energy Consumption Survey conducted by the Department of Energy. Examples of formulating questions and
activity logs are given in Chapter 5.

In addition to questionnaires and activity logs, much useful information can be
obtained by sketching the floor plans of buildings that are monitored and noting
the locations of doors, windows, appliances, and other contents. This approach
is practical if a technician visits the building to deploy and/or retrieve
monitoring equipment. Such a sketch provides useful information on building
volume, possible sources and sinks, and potential air movement patterns; moreover,
this information can be obtained with little or no disruption to the activities
of building occupants. Pictures of the building exterior and/or interior are
also useful records, provided that permission from occupants is obtained.

Evaluation of Efficacy

Evaluation of efficacy provides an important feedback loop for the systematic
development of survey instruments. The efficacy of survey instruments can be
assessed from three perspectives:

1. How much of a time burden is placed on the respondents for reporting or
    recording the desired information?

2. How accurate and complete is the reported and recorded information?

3. How does the information help interpret the monitoring results?

It is difficult to strike a balance between comprehensive questionnaires that may
overburden respondents and shorter questionnaires that may miss vital information.
To the extent practical, the respondent burden associated with survey instruments

should be minimized. In extreme cases, excessive reporting requirements could result in low participation rates or high attrition rates. For some questions, however desirable, respondents may be unable to provide accurate or complete information. The respondent's burden can be lessened somewhat by spreading questions across the three stages of monitoring and by using observations of technicians wherever possible.

To assess accuracy, quality-control or consistency checks can be made in different ways for static versus dynamic factors. For example, for some static factors, the technicians' observations during monitoring can be compared with occupants' descriptions that were obtained before monitoring. In some cases, independent data sources can be used such as tax assessment records to determine the age or size of a structure. For some dynamic factors, automated sensors can be used to complement or check an occupant's records of real-time practices. For example, a thermocouple or thermistor can be placed in the vicinity of a heat-producing appliance to detect periods of appliance activity. A pilot test can help to point out specific types of questions or components of activity logs with which respondents are having difficulty.

The contribution of specific information items to the interpretation of monitoring results can be assessed in a number of ways; one method for which objective criteria can be formulated is regression analysis. This method is generally applicable to cross-sectional monitoring studies because the dependent variable, pollutant concentration, is measured on an interval scale and the independent variables constructed from survey instruments typically combine interval and categorical scales. Various criteria can be used to assess the contribution of a specific variable such as the extent of increase in explained variance or the level of significance of a regression coefficient (Koontz and Nagda 1985).

Other multivariate methods that can be appropriate in specific instances are discriminant analysis, which is similar to regression analysis except that the dependent variable is measured by a few discrete categories, rather than on a continuum; factor analysis and principal component analysis, which seek to find related groups of factors or "dimensions" that may exist within a data set; and cluster analysis, which seeks to identify homogeneous groups or clusters of sampling units (e.g., structures) on the basis of factors other than those that are inherent in the sampling design. A full description of such procedures is beyond the scope of this book; for a more detailed treatment, the reader is referred to statistical texts (e.g., Draper and Smith 1966; Armitage 1971).

REFERENCES

American Society for Testing and Materials (ASTM), E779-81, Standard Practice for Measuring Air Leakage by the Fan Pressurization Method, ASTM, Philadelphia, 1981a.

American Society for Testing and Materials (ASTM), E741-80, Standard Practice for Measuring Air Leakage by the Tracer Dilution Method, ASTM, Philadelphia, 1981b.

Armitage, P., Statistical Methods in Medical Research, Blackwell Scientific Publications, London, 1971.

Card, W.H., Sallman, A., Graham, R.W., Drucker, E.E., Infrasonic Measurement of Building Leakage: A Progress Report in Building Air Change Rate and Infiltration Measurements (pp. 73-88), ASTM STP 719, C.M. Hunt, J.C. King, and H.R. Trechsel, eds., American Society for Testing and Materials, 1980.

Condon, P.E., Grimsrud, D.E., Sherman, M.H., and Kammerud, R.C., An Automated Controlled-Flow Air Infiltration Measurement System, LBL-6849, Lawrence Berkeley Laboratory, Berkeley, 1978.

Conway, J., Real World Interfaces, Comput. Electron., vol. 22, pp. 54-58 and 104-106, 1984.

Cooney, T.M., Portable Data Collectors, And How They're Becoming Useful, J. For., pp. 18-23, January 1985.

Dietz, R.N., and Cote, E.A., Air Infiltration Measurements in a Home Using a Convenient Perfluorocarbon Tracer Technique, BNL 30797R, Brookhaven National Laboratory, Upton, New York, 1982a.

Dietz, R.N., and Cote, E.A., An Inexpensive Perfluorocarbon Tracer Technique for Wide-Scale Infiltration Measurements in Homes, Environ. Int., vol. 8, pp. 419-433, 1982b.

Dietz, R.N., D'Ottavio, T.W., and Goodrich, R.W., Multizone Infiltration Measurements in Homes and Buildings Using a Passive Perfluorocarbon Method. ASHRAE Trans., vol. 91, part 2, paper HI-85-35, no. 3, 1985.

Dillman, D.A., Mail and Telephone Surveys, the Total Design Method, John Wiley and Sons, New York, 1978.

Draper, N.R. and Smith, H., Applied Regression Analysis, John Wiley and Sons, New York, 1966.

Fitz-Simmons, T., and Sauls, H.B., Using the HP-41CV Calculator as a Data Acquisition System for Personal Carbon Monoxide Exposure Monitors, J. Air Pollut. Control Assoc., vol. 34, pp. 954-956, 1984.

Graham, R.W., Infrasonic Impedance Measurement of Buildings for Air Leakage Determinations, Technical Report TR-77-15, Department of Electrical Engineering, Syracuse University, Syracuse, New York, 1977.

Grimsrud, D.T., Sherman, M.H., Hanssen, J.E., Pearman, A.N., and Harrje, D.T., An Intercomparison of Tracer Gases Used for Air Infiltration Measurements, ASHRAE Trans., vol. 86, no. 1, pp. 258-267, Atlanta, 1980.

Harrje, D.T., Gadsby, K., and Linteris, G., Sampling for Air Exchange Rates in a Variety of Buildings, ASHRAE Trans., vol. 88, p. 1, Atlanta, 1982.

Harrje, D.T., Dutt, G.S., Bohac, D.L., Gadsby, K.J., Documenting Air Movements in Multicell Buildings Using Various Tracer Gas Techniques, ASHRAE Trans., vol. 91, part 2, paper HI-85-40, no. 3, 1985.

Hernandez, T.L., and Ring, J.W., Indoor Radon Source Fluxes: Experimental Tests of a Two-Chamber Model, Environ. Int., vol. 8, pp. 45-57, 1982.

Koontz, M.D., and Nagda, N.L., Systematic Development of Survey Instruments for Indoor Air Quality Studies, Proceedings of the 78th Annual Meeting of the Air Pollution Control Association, Pittsburgh, paper no. 85-31.2, 1985.

Lagus, P., and Persily, A.K., A Review of Tracer-Gas Techniques for Measuring Airflows in Buildings, ASHRAE Trans., vol. 91, part 2, paper HI-85-22, no. 1, 1985.

Liddament, M., and Thompson, C., Techniques and Instrumentation for the Measurement of Infiltration in Buildings--A Brief Review and Annotated Bibliography, Technical Note AIC-TN-10-83, Air Infiltration Centre, Berkshire, Great Britain, 1983.

Mintz, S., Hosein, H.R., Batten, B., and Silverman, F., A Personal Sampler for Three Respiratory Irritants, J. Air Pollut. Control Assoc., vol. 32, pp. 1068-1069, 1982.

Modera, M.P., and Sherman, M.H., AC Pressurization: A Technique for Measuring Leakage Area in Residential Buildings, ASHRAE Trans., vol. 91, part 2, paper HI-85-03, no. 3, 1985.

Nagda, N.L., Koontz, M.D., Fortmann, R.C., and Rector, H.E., Air Infiltration and Building Factors: Comparison of Measurement Methods, Proceedings of the 79th Annual Meeting of the Air Pollution Control Association, Pittsburgh, 1986.

Sandberg, M., The Multi-Chamber Theory Reconsidered from the Viewpoint of Air Quality Studies, Build. Environ., vol. 19, no. 4, pp. 221-233, 1984.

Sherman, M.H., and Grimsrud, D.T., Measurement of Infiltration Using Fan Pressurization and Weather Data, LBL-10852, Lawrence Berkeley Laboratory, Berkeley, 1980.

Sinden, F.W., Multi-Chamber Theory of Infiltration, Build. Environ., vol. 13, pp. 21-28, 1978.

Chapter 5
DESIGN

Development of a design for indoor air quality monitoring requires systematic thinking along many, sometimes conflicting, avenues. This thinking should address what pollutant(s) to monitor and what instrument(s) to use. Further considerations include how many locations to monitor within a given building, how long to monitor, and how many buildings to monitor. Attention should also be given to identifying other measurable factors that modify indoor concentrations such as air exchange. A hastily developed design could very well lead to results of little practical significance or, worse, to erroneous and misleading findings. Thus, good design requires a considerable amount of thought and effort.

Programs for indoor air quality typically are divided between supporting applied research and investigating building-associated problems. The major difference between these two study types lies in the ability to safely know ahead of time what to measure.

Applied research programs are undertaken to establish relationships between indoor pollutant concentrations and one or more mass balance factors; examples include measuring the impact of emissions from specific indoor sources or evaluating the indoor air quality consequences of large-scale weatherization programs. Building investigations are undertaken to discover which pollutants and mass balance factors apply to a situation initially prompted by health complaints.

Design principles for applied research programs are discussed in two stages: development of a preliminary research plan followed by development of a detailed monitoring design. Because investigations of building-associated problems have been carried out in the past on an ad hoc basis by investigators with varied backgrounds, common methods and well-defined protocols for such investigations are not available. A separate section of this chapter describes avenues to be pursued in addressing building-associated indoor air quality problems based on the current (late-1985) state of the art.

DEVELOPMENT OF PRELIMINARY RESEARCH PLAN

The basic input to the development of a preliminary research plan is the overall study objective or the reason for monitoring. The study objective may be supplied by an organization or individuals other than those who will conduct the actual monitoring. The output of the process is a preliminary research plan, which becomes an input to the development of the detailed monitoring design.

The development of a preliminary research plan is an iterative process involving the three steps shown in Figure 12. A discussion of each step follows.

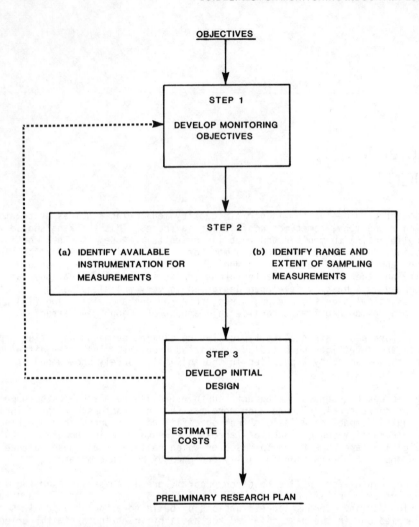

FIGURE 12. Development of initial design for indoor monitoring.

Step 1--Develop Monitoring Objectives

Developing a clear statement of monitoring objectives from study objectives is
the first critical step in the design process. Monitoring objectives state what
are the pollutants to be measured, what is the relative importance of pollutants
to be monitored if there is more than one pollutant, what are other factors to be
measured, and what are some possible design alternatives that should be considered.

At first glance, this step seems trivial because most study problems can be
considered synonymous with objectives. However, it is essential that these objectives be expanded into monitoring objectives. The process of developing monitoring objectives involves a systematic interrogation of the problem setting, the

problem background, and the knowledge base needed to form both qualitative and quantitative goals that are as specific as possible.

Some typical study objectives are as follows:

● To characterize the range and distribution of pollutant concentrations in certain types of structures

● To determine the proportion of total pollutant exposure that is attributable to indoor exposure

● To assess the effect of building tightness on indoor air quality

● To identify factors that are associated with indoor air quality problems in residences or other buildings

● To test indoor air pollutant mitigation measures.

Monitoring objectives can be developed from such study objectives by addressing the following questions:

A. What pollutants are important to measure?

B. What other types of factors, in addition to the pollutants of concern, should be measured to satisfy the monitoring objectives?

C. What are the important monitoring considerations and alternatives?

D. What questions cannot be addressed by the monitoring effort?

If the monitoring objectives stipulate more than one pollutant of concern, then the relative importance of pollutants should also be addressed. In some studies, such as the investigation of indoor air quality problems tied to occupant complaints, the list of candidate pollutants may be quite broad. Monitoring approaches for this type of objective are discussed separately in the last section of this chapter.

To determine other types of factors for which measurements may be needed to satisfy study objectives, the terms in the mass balance equation are described in Chapter 3. In addition to indoor concentrations, some of the mass balance terms that determine these concentrations can be measured directly--C_{out} (outdoor concentration), ν (air exchange rate), and V (volume of the structure). The terms S (source generation rate), λ (removal and decay rate), F_B (filtration), and m (mixing) cannot be measured directly and require special experimental provisions, such as chamber studies for S and λ or modeling of full-scale indoor environments. The illustrative set of related factors listed in Table 4 of Chapter 3 can be reviewed for applicable terms from the mass balance equation. As described in Chapter 4, instrumentation exists to measure factors such as air exchange rates and indoor/outdoor environmental variables.

Table 13 illustrates the application of questions A through D above to three examples. The study objectives and scopes of the three cases are quite varied.

For example 1, regional weatherization, radon (Rn) is the most important pollutant to be monitored, followed by formaldehyde (HCHO) and inhalable particulate matter (IP). IP is important if the weatherization program is in an area where wood is used fairly often as a heating fuel or if tobacco smoking is to be considered in the design. Rn progeny is of lesser importance because it is highly dependent on

TABLE 13. Examples of Issues to be Addressed in Developing Monitoring Objectives from Study Objectives

Study objectives	Monitoring objectives
Example 1. Regional weatherization	
Assess the indoor air quality impacts attributable to a statewide residential weatherization program	A. Pollutants: radon, radon progeny, formaldehyde, inhalable particulates, carbon monoxide, nitrogen dioxide, and organics.
	B. Other factors: air exchange, building tightness, temperature, humidity, outdoor concentrations.
	C. Considerations and alternatives: effects may vary with a number of factors such as season, geology, construction materials, interior furnishings, and occupant activities. Monitoring strategies include pre-/postweatherization contrasts or monitoring separate groups of weatherized and unweatherized houses.
	D. Questions that cannot be answered: detailed assessments for individual houses will be difficult unless each house is monitored year round.
Example 2. Source testing	
Determine potential indoor air quality impacts of an advanced burner design for an unvented, fossil-fuel-fired space heater	A. Pollutants: carbon monoxide, nitrogen dioxide, inhalable particulates, organics, and possibly sulfur dioxide.
	B. Other factors: air exchange, temperature, humidity, fuel analysis, fuel consumption, energy balance of test environment.
	C. Considerations and alternatives: pollutant emissions of concern will depend upon fuel type. Given a limited number of prototypes, testing opportunities may be restricted to chamber environments and/or a few typical indoor environments.
	D. Questions that cannot be answered: how occupant usage patterns, maintenance, and details of the indoor environment, such as furnishings and floor plan, may affect performance and pollutant emissions.
Example 3. Radon mitigation	
Evaluate alternative strategies for radon mitigation in residences	A. Pollutants: radon and radon progeny.
	B. Other factors: air exchange, outdoor concentrations, suitability of specific mitigation strategies for specific housing types.
	C. Considerations and alternatives: finding representative houses is important; radon entry varies tremendously with geology and construction details. Nonetheless, monitoring a large number of houses will facilitate generalization but will also restrict the level of experimental control.
	D. Questions that cannot be answered: cost effectiveness of alternative strategies for all types of homes.

Rn gas concentration, although its measurement is useful. Nitrogen dioxide (NO_2) and carbon monoxide (CO) could be of importance if unvented combustion sources are present indoors.

For example 2, source testing, both CO and NO_2 are expected to be important because they are normally products of fossil-fuel combustion. Depending on the particular type of fuel, pollutants such as sulfur dioxide (SO_2), IP, volatile organic compounds, and polynuclear aromatics may be of importance.

For example 3, radon mitigation, both Rn and Rn progeny are of importance because some mitigation strategies can reduce Rn progeny levels by reducing Rn concentrations, whereas others can reduce Rn progeny levels without affecting Rn.

Air exchange is probably the single most important factor after pollutants for measurement in any indoor air quality study. Indeed, it would be more difficult to define a study that would not benefit from air exchange data than to figure out reasons for excluding this factor. With information on air exchange rates, data interpretation is strengthened and results can be projected to other conditions.

In each of the examples cited here, the fundamental objective is to define the consequences of a change in the mass balance. In the regional weatherization example, this change is primarily in the air exchange rate. In the source testing example, the change is primarily in the source term. In the radon mitigation example, the changes may be in the source term as well as in special removal terms, depending on the mitigation strategy. Using the mass balance to interpret mitigation effectiveness is particularly important in this type of study because the natural variability of the Rn entry rate can be a confounding factor.

House characteristics and occupant activities are important for regional weatherization because the effects of weatherization are dependent on these factors. A limited set of house characteristics, particularly those relating to type of foundation, are important to radon mitigation. Occupant activities that relate to air circulation and ventilation are also important confounding factors for this example; consequently, it may be best to monitor only a limited number of unoccupied houses. This approach will afford a much greater degree of experimental control but will also limit the degree to which results can be generalized to other types of houses and to occupied conditions. For the source testing example, factors such as temperature, humidity, fuel consumption, and energy balance are important because they are associated with heater performance.

Because monitoring resources are usually restricted, the realistic scope and limitations of the study should be considered at this early design stage. A monitoring design that is too ambitious in scope may dilute available resources and thereby fall short of addressing the fundamental monitoring objective. However, it is also quite reasonable to expect that a monitoring study can address more than one objective. If a study has multiple objectives, then these should be clearly ranked so that the more important goals are not compromised as the design develops.

In this first step, monitoring objectives are formulated. The outputs of this step should be statements addressing questions A, B, C, and D as outlined in Table 13. In developing monitoring objectives, not all questions can be fully articulated for various reasons including lack of specificity in study objectives and lack of sufficient knowledge. Thus, the monitoring objectives will likely become further refined in later stages, but maintaining a written statement will help reduce confusion that could be created because of a large number of choices available.

Step 2a--Identify Available Instrumentation for Measurements

Instrumentation for indoor measurements includes both monitoring devices and survey instruments. Monitoring devices either provide a series of measurements that can be recorded, usually by electronic means, or provide a sample that is later analyzed, usually in a laboratory setting. Survey instruments require manual methods to record physical properties of a structure or activities that take place within it. Some properties, such as the volume of a structure, may require physical measurements whereas attributes, such as the presence of a specific type of appliance, can be determined by observation.

For each measurement factor under consideration, various aspects of available measurement systems should be reviewed. The information obtained on each piece of instrumentation need not be highly detailed at this point, but should cover the following points:

- Is an off-the-shelf device available for conducting measurements or is the application of a method required, as outlined in Chapter 4?

- Is the instrument portable?

- Does the instrument require a power source to operate?

- Does the instrument produce a real-time measurement or does it collect a sample that requires subsequent analysis?

- Does the instrument provide only a summary, integrated measurement or does it produce a continuous output signal?

- Must the instrument be in place for a minimum length of time to produce a measurement of reasonable quality?

- Over what range of concentrations can the instrument provide measurements of acceptable quality?

- Is the performance of the instrument dependent on any specific factors such as temperature, humidity, or interferents?

- What is the cost of the instrument?

Examples of this type of information are shown in Table 14 for three alternatives for NO_2 measurements. The first two are monitoring devices that produce continuous signals and the third is a method by which passive samplers are exposed to collect an integrated sample for laboratory analysis. One of the continuous devices is lighter and thus more easily transported from place to place. The portable unit also consumes less power. The concentrations measured by the passive samplers vary slightly with temperature, but the nature of this relationship has been quantified. These samplers require a fairly long exposure period (nominally, a week) to accumulate sufficient sample material. A summary of instrument characteristics similar to Table 14 can be assembled for other pollutants.

TABLE 14. An Example of Characteristics Associated with Three Different Measurement Approaches for NO$_2$

Characteristics	Stationary analyzer	Portable analyzer	Passive collector
Type of device or method	Monitoring device purchased from manufacturer	Monitoring device purchased from manufacturer	Samplers can be purchased or fabricated and recycled
Portability	Weighs approximately 34 kg	Weighs about 10 kg; can be moved from house to house	Very small and lightweight; can be used for personal monitoring
Power source	120 V ac	Battery with electrical backup; 120 V ac; 12 V dc battery	None required; operates by diffusion
Measurement output	Continuous signal	Continuous signal	Integrated sample requiring laboratory analysis
Time limitations	Essentially no limitations	Essentially no limitations	Depends on concentration; should have at least 1 ppm h of exposure
Concentration range	0.002 to 5 ppm	0.010 to 5 ppm	1 to 20 ppm h (ultimate sorbent capacity exceeds 1 000 ppm h)
Dependency on other factors	Little to none	Little to none	Slight dependency on temperature
Approximate costs	$7 000 per unit	$7 000 per unit	$10 per tube, $20 per laboratory analysis at commercial rates; lower if fabricated and analyzed in-house

A wide variety of available instruments and methods with varying degrees of sophistication and associated costs are reviewed in Chapter 4 and in the appendixes. New instrumentation--especially personal and portable devices--is continuously being developed, tested, and marketed. Thus, the summary of available instrumentation and methods contained here may need to be updated before an initial design is developed. Supplementary information can be obtained in a number of ways such as:

- By consulting with sponsoring and performing organizations that are active in the field of indoor air quality; many such organizations are referenced throughout the text.

- By locating review articles that describe the state of the art of monitoring equipment; some examples are cited in Chapter 7.

But at this early stage it may suffice to use instrumentation information given
for this book or information generally available rather than a full-scale
instrumentation review that may divert attention. Actual selection of instrumentation is a step in the detailed design, described later. A different type of
instrument that should be considered at this stage of the monitoring design is
the survey instrument, or questionnaire. Such an instrument is useful for qualitatively or quantitatively describing certain characteristics of indoor environments or occupant practices that cannot be measured. Such descriptions can often
enhance the interpretation of monitoring results. In concert with the identification and cursory review of monitoring devices, consideration can be given to
the nature and types of the survey instruments that will be required. This
thinking may be aided by reviewing survey instruments that have been used in
previous studies.

Step 2b--Define Range and Extent of Sampling Requirements

An evaluation of the approximate range of sample size early in the design plan
can be useful in selecting instrumentation and in determining the approximate
extent of monitoring required. Thus, this design step should be performed in
parallel with the initial review of instrumentation, as depicted in Figure 12.
The emphasis here is on range of sample size, not the actual sample size, which
is discussed later.

The range of sample size is dependent on objectives. For example, if the main
purpose of monitoring is to develop models for seasonal, time-varying concentrations of various pollutants, then the number of houses may be very limited--
even as few as one or two. This approach will permit extensive measurements of
various pollutant concentrations and air exchange rates as inputs to model
formulation and testing. For this type of study, stationary instruments with
active analytical devices are usually the most suitable.

The opposite extreme is a study that investigates whether a particular type of
problem exists, for example, high Rn concentrations in houses in a particular
locality. Such determinations may need to be based on a cross-section of houses.
For a representative sample of several hundred or more houses, less expensive
passive instruments might be used. Of course, the concentrations obtained with
passive monitoring will be averaged over the duration of the monitoring period,
with no identification of short-term peaks.

Also to be considered under the range of sample sizes are the geographic loca-
tion(s) for monitoring and the type(s) of indoor environments that need to be
addressed. Often the location may be stated in the objectives (e.g., assessment
of the impact of weatherization on indoor air quality in residences in a specific
region of the country). But in some cases, there is flexibility in the selection
of appropriate location(s). Heating and cooling degree days, outdoor pollution
levels, population density, and patterns of building design and construction are
among the factors that should be considered in selecting geographic locations.

Table 15 illustrates the application of these considerations to the three example
study objectives that were used for step 1. In the regional weatherization
example, the potential sample size to represent a statewide area is quite large,
as many as 1 000 units; this large sample size may be required because of the
many factors that can confound the effects of weatherization. For the source
testing example, the number of heater prototypes may be limited; therefore, the
opportunities for testing in different structures is necessarily restricted.
Moreover, a high degree of experimental control may be desirable to compare
contaminant levels resulting from the new burner design with those from an

existing design. In practical terms, the strategy of choice is linked to general
properties of a given home rather than specific details. Therefore, use of
"typical homes" to represent generic classes is appropriate. In example 3,
either "typical" single-family homes or a broader cross-section of homes could be
used to test mitigation approaches, but the most important point is that each
home treated for mitigation is to be paired with a similar but untreated home for
comparison purposes.

**TABLE 15. Three Examples of Range and Extent of Sampling Requirements for
Different Study Objectives**

Objectives and requirements	Example 1-- Regional weatherization	Example 2-- Source testing	Example 3-- Radon mitigation
Study objectives	Assess the indoor air quality impacts attributable to a statewide residential weatherization program	Determine potential indoor air quality impacts of an advanced burner design for an unvented, fossil-fuel-fired space heater	Evaluate alternative strategies for radon mitigation in residences
Range of sample size	200 to 1 000 units; summer and winter seasons	1 to 4 units; winter season	2 to 200 units; all seasons
Types of indoor environments	Single-family houses with different foundations, ages, construction materials, and heating fuels	Test chamber, "typical" single-family homes or apartments	Pairs of single-family homes with concrete slab, crawl space foundations; one of each pair to be treated for mitigation and the other to be used for comparison purposes
Range of geographic location	Various parts of the State, depending on geologic profile; should include both urban and rural settings	A single geographic location may be sufficient	An area where relatively high radon concentrations have been measured

For the regional weatherization example, periods of warm weather are important
because HCHO concentrations should be highest at these times. Winter is important
to assess effects associated with weatherization and emissions from heating
appliances. To characterize the annual distribution of Rn/Rn progeny concentrations, measurements are desirable for all seasons. For the source testing example,
winter is the season of major concern; for radon mitigation, all seasons should
be considered. Various types of indoor environments and geographic locations

need to be considered for a regional-scale study, but these factors can be deliberately restricted without compromising the study objectives for source testing.

In summary, the operative word for step 2 of the monitoring design is to identify. The intent of this step is to become aware of the range of alternatives for instruments, sample size, and geographic location. Some tentative choices can be made at this point, but final decisions should be reserved for subsequent steps in the design process. It may be readily apparent at this point that certain combinations of alternatives, though attractive from a technical standpoint, are not practical. For example, it may be desirable to monitor 500 houses during different seasons in 10 geographic locations using only devices that produce continuous signals, but this particular alternative requires substantial equipment and labor resources and thus carries a hefty price tag.

Step 3--Develop Initial Design

The development of monitoring objectives under step 1 and the identification of instrumentation and sampling requirements under step 2 are fundamental inputs to the initial design development (step 3). In this step, all issues are integrated and simultaneously considered by addressing the following types of questions:

● What type of and how many structures are involved?

● What types of activities occur in these structures?

● Over what period of time is the measurement for each pollutant to be taken?

● When should monitoring take place?

● Will monitoring occur on all days of the week or only on selected days?

● Can average pollutant concentrations be monitored with collectors (passive or active) or must peak concentrations be measured?

● If collectors are not adequate, will intermittent monitoring be sufficient, or is continuous monitoring required to meet objectives?

● How many monitoring locations per pollutant and per structure are required?

● Are outdoor measurements also needed for each pollutant?

● What factors other than pollutants will be measured, and how will measurement of these factors be integrated with pollutant monitoring?

● Will measurements for different structures be completely independent, or will they parallel one another in some systematic way?

Monitoring frequency, duration, and location are partially dictated by study objectives and instrumentation preferences. For example, a study of the effects of structural tightness on particulate matter levels during wood stove operation will focus on the heating season. The types of activity and patterns of building use are also important. Passive monitoring studies will require a sufficient monitoring time to ensure that minimum detection levels are exceeded. Studies of the effect of traffic patterns on residential CO levels might be restricted to certain hours of the day.

An additional consideration is the manner in which the initial design meets the sample size requirement. This requirement can be met in two ways: one is by selecting many units (e.g., buildings) and sampling each one for short periods of time (e.g., a day or a week), and the other is by selecting only a few units and sampling each for longer periods (e.g., a season or a year). These two approaches are not usually equivalent. In experimental situations, the latter option often must be pursued. Otherwise, some compromise between the two extremes may be preferred. The use of a very limited number of houses can afford a higher degree of experimental control and can yield a more detailed understanding of pollutant behavior and associated physical or chemical processes, but the extent to which monitoring results can be generalized to other buildings will be restricted.

Development of the initial design will benefit from anticipating the types of analysis that will be performed when the monitoring is completed. For example, if the data are intended to support modeling efforts or other forms of time-series analysis, then continuous measurements are needed. It must be determined whether the monitoring is exploratory in nature or whether statistical inferences are to be made based on the sample of structures that are monitored. Statistical analysis will address one or more of the following:

- Parameters describing the temporal or cross-sectional distributions of indoor pollutants (e.g., mean, standard deviation, percentiles)

- Hypotheses concerning differences among various sampling conditions (e.g., different types of structures or structures with different contents)

- Hypotheses concerning relationships between various factors and pollutants.

The monitoring objectives should be reexamined at this point to determine if they can be restated in the form of hypotheses to be tested or as parameters to be estimated.

The output of the initial design development is a preliminary research plan that addresses the following points:

- The general background and purpose of the monitoring effort, recognizing whether the monitoring objectives constitute exploratory research or imply specific estimation or hypothesis-testing goals

- A list of pollutants and other factors for which monitoring is desired and a tentative choice of one or more instruments for each

- A monitoring timeframe (e.g., approximate duration, seasons of interest)

- A target sample size or size range and a tentative list of study locations or criteria for choosing locations

- Definition of the sampling units (e.g., structures) for monitoring and a general description of procedures for identifying and selecting these units

- A description of the temporal and spatial detail sought for measurements within each sampling unit

- Comments on types of data analyses to be performed

- A tentative schedule for activities before monitoring, during monitoring, and after monitoring.

Generally, this plan should be about 5 to 15 pages long and should be in a format suitable for review by management, by the sponsor, and by peers, as appropriate. A preliminary estimate of costs is prepared at this stage but can be separated from the document containing research plans. The plan developed based on the above points and within the specified range of page length will contain sufficient information to convey the rationale for monitoring and the overall design and scope of the monitoring effort. Such a concise plan without excessive details will likely result in a timely review. Perhaps a research plan may also entice helpful suggestions from the reviewers, which will strengthen the design. Unfortunately, research plans are generally not published; however, examples of research plans may be found in Nagda et al. (1983, 1984) and Fortmann and Nagda (1985).

DEVELOPMENT OF DETAILED MONITORING DESIGN

Figure 13 shows six additional steps that transform the preliminary research plan into a detailed monitoring design. Two important issues--measurement procedures and sample-selection procedures--are treated in parallel. These issues are later merged in a set of protocols that are subjected to a pretest and refined on the basis of the pretest experience. Cost estimates are prepared in step 9 based on information gathered in prior steps. The ultimate product is a detailed design that conveys the overall scope of the monitoring, specific instruments and procedures, and associated costs.

Step 4--Select Instrumentation for Measurements

The operative word for this design step is to select, rather than simply identify or review instrumentation. In general, the selection will be guided by the objectives, by initial sample size estimates, by the degree of spatial and temporal detail required, and by available resources.

Monitoring devices. To select monitoring devices, it is important to have some idea of the range of values that may be encountered and their temporal cycling frequency (i.e., how rapidly can values rise or fall over time and over what range?). For example, even if pollutant instruments that provide continuous signals are used, response to rapidly changing concentrations may lag. Similarly, because it is normally not practical to record or analyze the instantaneous response of every instrument, some averaging time is chosen (e.g., 15 min, 1 h). If the averaging period is too short, recording capacities may be exceeded or the volume of data to be analyzed may be overwhelming. If the averaging period is too long, important temporal variations may be masked and the variance of the pollutant may be underestimated (Saltzman 1970). The choice of averaging period may be driven by other considerations such as the averaging period for which air quality standards have been stated; some averaging periods used for short-term ambient air quality standards are 1, 8, and 24 h.

In addition to factors listed under step 2 that should be reexamined here, further considerations for the selection of monitoring devices are as follows:

● Mobility--the monitoring approach could be strictly limited to one class of instrument mobility (i.e., stationary, portable, or personal), but it is sometimes possible to use two or even all three classes effectively.

● Size--some devices, although attractive from the standpoint of performance, may be too large or noisy for placement in an occupied environment; a possible alternative is to place a sampling line in the environment to be monitored and to locate the device in a separate environment.

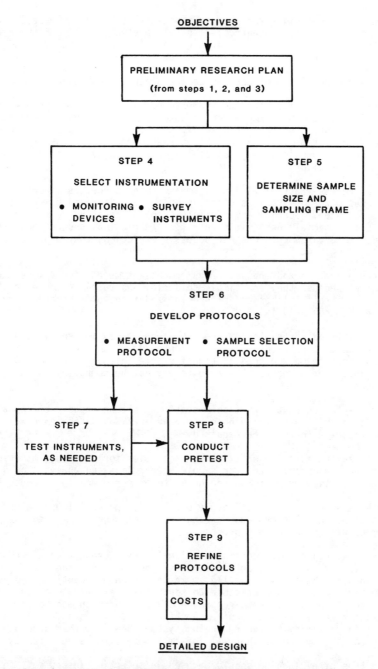

FIGURE 13. Development of detailed design for indoor monitoring.

- Accuracy and precision--accuracy is a measure of how close a device comes to measuring the actual concentration that is present in the monitored environment; precision is a measure of the extent of agreement between two devices of the same type that are monitoring the same environment. There is usually a trade-off between the size of a device and its accuracy or precision; the accuracy and precision of any device are often concentration dependent.

- Calibration and maintenance requirements--some devices can operate unattended for periods of a week or more whereas other devices require frequent maintenance or adjustments; the response of instruments to a specific concentration can drift over time in a random or systematic manner; if the drift is large in magnitude or unpredictable in direction, then the device may need to be checked or recalibrated frequently.

- Appropriateness--factors within the measurement scene are important. Occupied residences place greater restrictions on size, noise, and power than do unoccupied structures. Special environments such as operating rooms and classrooms may require special intrinsic safety (fire/explosion) precautions. All types of occupied environments present access restrictions that must be balanced against maintenance and calibration requirements.

- Sample preparation and analysis requirements--some passive devices may appear attractive because they are simple to deploy and retrieve; however, they may require substantial labor and/or specialized types of laboratory equipment for preparation and analysis.

- Availability and costs--the desired instruments may currently be on hand or may have to be purchased, leased, or borrowed; if the devices are not on hand, they may need to be acquired in time for some initial testing before monitoring begins; and, moreover, their costs may not be compatible with available resources.

- Personnel--the currently available staff may not be capable of operating all devices; in this case, training of current staff, addition of new staff, or use of a contractor will be required.

- Facilities--the adequacy of existing facilities for equipment maintenance, repair, and calibration and for sample preparation and analysis needs to be examined, particularly if monitoring is to be performed in a remote location.

- Permanence--equipment selection may be based on potential uses after monitoring is complete; loan or lease arrangements for short-term needs of permanent equipment will need to be explored.

If one or more devices with continuous signals are chosen, then a method of recording these signals must be selected; various types of recording devices and systems are described in Chapter 4. Ideally, when more than one continuous device is used within a structure, it is preferable to route their signals to a central recording system rather than use separate recording devices for each monitor. This approach may lead to excessive wiring requirements for signal lines unless all devices can be located in proximity to the recording system; however, the advantage of a central recording system is that all information has the same time base.

This same type of evaluation should be used in selecting instruments for additional factors such as air exchange rate and environmental variables. In some areas, meteorological data collected by the National Weather Service may be a viable substitute for onsite anemometry, thermometry, and hygrometry.

Survey instruments. As noted earlier, survey instruments are an important supplement to the monitoring process. In contrast to monitoring devices whose operating characteristics are not easily changed, survey instruments are usually unique
to each monitoring study. The development of survey instruments, which essentially are a few pieces of paper, may not appear to require substantial resources;
however, a careful development will require a significant amount of researchers'
time and efforts. Similarly, the instruments' implementation can have a sizable
impact on the resources required during monitoring and during subsequent stages
of data processing and analysis.

Some general considerations for formulating survey instruments are outlined in
Chapter 4. The following specific steps can be applied:

● Identify the number of survey instruments to be used and the following
  elements for each:
  - The general role of the instrument in relation to the monitoring effort
  - The individual who will administer the instrument
  - The stage of monitoring during which the instrument will be administered
  - The extent of interaction with building occupants required to complete
    the instrument

● Determine the factors to be characterized with survey instruments by reviewing
  the mass balance terms and associated variables listed in Table 4 of Chapter 3

● Assign each factor to be characterized to the most appropriate survey instrument, if more than one instrument is to be used.

Methods of characterizing static and dynamic factors are summarized in Table 12
of Chapter 4. An example application of these methods to characterize building
tightness and air exchange is given in Table 16. For static factors relating to
building tightness, several questions are posed to occupants at the premonitoring
stage, some observations are made by technicians during monitoring, and a blower
door measurement is taken after monitoring. For dynamic factors related to air
exchange, a question about usual occupant practices is posed prior to monitoring.
During monitoring, actual practices are recorded on an activity log; as an
alternative, actual practices are determined retrospectively after monitoring has
been completed. A measurement of the prevailing air exchange rate is also taken
concurrently with air pollution monitoring.

**TABLE 16. An Example of Questions and Measurements to Characterize Building
Tightness and Air Exchange for an Indoor Air Quality Study, Classified by Nature
of Factor and Stage of Monitoring**

Stage of monitoring	Nature of factor requiring characterization	
	Static	Dynamic
Premonitoring	Questions to occupants about factors related to building tightness	Question to occupants about their usual frequency of opening windows
Monitoring	Technician observations about potential exterior barriers to wind forces	Activity log* to denote times when windows were opened or closed; air exchange measurement (tracer gas)
Postmonitoring	Blower door measurement	Question* to occupants about general window-opening practices during the monitoring period

* Generally one of these two alternatives would be sufficient.

To illustrate the types and wording of questions for static factors related to air exchange, some sample questions are formulated in Figure 14. Questions posed during premonitoring concern the age of the building (question 1), specific modifications intended to increase building tightness (questions 2 and 3), and perceived draftiness of the building under prescribed conditions (question 4). Technician observations during monitoring include nearby trees or structures that might alter wind exposure (question 1), building location relative to surrounding terrain (question 2), and a judgment concerning the relative exposure of the building (question 3). The three questions collectively can address potential wind impacts on the air infiltration rate.

Sample questions for dynamic factors related to air exchange are given in Figure 15. Because such factors vary by definition, technician observations cannot be used here; all information must be supplied by the building occupants or monitored through automated devices. The premonitoring question addresses usual practices concerning window openings, the activity log addresses actual practices during monitoring, and the postmonitoring question seeks to obtain information similar to that recorded on the activity log, but in less detail. Generally, either the activity log or the postmonitoring question, but not both, would be used.

The above framework and alternatives for development of survey instruments are given to illustrate an application of methodology given earlier. The exact content and form of survey instruments depend on the monitoring and study objectives.

The specific questions for each survey instrument need not be formulated at this point. However, as a minimum, the nature of information sought from each question should be identified and the flow of questions should be charted. The questions and wording used in survey instruments from previous studies may provide some useful guidance, but it is important to recognize that the study objectives ultimately affect the choice of questions to be included and the wording of these questions. Each item to be included in a survey instrument should have an eventual role in data analysis.

Step 5--Determine Sample Size and Sampling Frame

Sample size. In determining sample size, the preliminary selection of equipment and previously estimated range of sample size serve as useful starting points. In addition to specific objectives of the monitoring design, the sample size (i.e., the total number of air samples) depends on the following types of factors:

- Pollutant(s) to be monitored

- Type of structure(s) to be monitored

- Geographic area(s) where monitoring is to take place

- Season(s) of the year during which monitoring is to take place

- Day(s) of the week on which monitoring is to take place

- Length of the time interval during which each sample is taken (e.g., grab sample, 1-h sample, 24-h sample).

Estimates from previous studies of the average pollution levels and variation around this average will be helpful in estimating sample size. For the formulas presented below, preliminary estimates of the arithmetic mean, $\overline{X}$, and standard deviation, S, are required. The sampling conditions such as pollutant, structure

PREMONITORING (answered by occupants)

1. When was this building built? _____ [ ] (check if guessed)

2. How many of your doors and windows have storm doors and windows, double-glazed glass, or other protective coverings such as plastic or shutters?

   [ ] All or most   [ ] Some or about half   [ ] Few or none

3. During the past 2 years, has any caulking or weatherstripping been applied to your doors or windows?   [ ] Yes   [ ] No

   (If yes) For how many of the doors or windows was this done?

   [ ] All or most   [ ] Some or about half   [ ] Few or none

   What is the current condition of the caulking or weatherstripping?

   [ ] Basically intact   [ ] Cracked or peeling   [ ] Don't know

4. During times when it is windy or cold outside, do you notice or feel drafts around ....

   a. Doors?   [ ] Often   [ ] Sometimes   [ ] Rarely

   b. Windows?   [ ] Often   [ ] Sometimes   [ ] Rarely

   c. Any other place? (Specify) _____

- -

MONITORING (answered by technicians)

1. Are there any of the following within 50 feet of any side of the building? (check all that apply)

   a. Groups of evergreen trees
      at least 4 feet tall?   [ ] Front [ ] Rear [ ] Left*   [ ] Right*

   b. Groups of other trees
      as tall as the building
      or taller?   [ ] Front [ ] Rear [ ] Left   [ ] Right

   c. Buildings as tall as the
      building or taller?   [ ] Front [ ] Rear [ ] Left   [ ] Right

   * When facing the front of the building.

2. Which of the following best describes the location of the building?

   [ ] At the top of a hill   [ ] On the slope of a hill
   [ ] At the bottom of a hill   [ ] On a relatively flat area

3. Relative to other buildings in the area, is this building generally exposed to the wind ....

   [ ] More than others
   [ ] Less than others
   [ ] About the same as others

Figure 14. Sample questions on such static factors as building tightness and wind barriers.

PREMONITORING (answered by occupants)

1. At this time of the year, how often do you usually open windows to let
    outside air in?

    [ ] Very often   [ ] Often   [ ] Sometimes   [ ] Rarely   [ ] Never

MONITORING (activity log completed by occupants)

Day	Time of day	Which of the following did you do at this time? (check only one answer per line)	
_____	_____	a.m.* [ ] Opened one window p.m.* [ ] Shut some windows	[ ] Opened two or more windows [ ] Shut all windows
_____	_____	a.m. [ ] Opened one window p.m. [ ] Shut some windows	[ ] Opened two or more windows [ ] Shut all windows
_____	_____	a.m. [ ] Opened one window p.m. [ ] Shut some windows	[ ] Opened two or more windows [ ] Shut all windows
_____	_____	a.m. [ ] Opened one window p.m. [ ] Shut some windows	[ ] Opened two or more windows [ ] Shut all windows

*Circle a.m. or p.m. each time that you enter the time of day.

POSTMONITORING (questions posed by a technician to occupants)

1. How often did you open windows during the period when we monitored your
    building? (Describe in as much detail as the occupant provides.)

    _____

    _____

    _____

FIGURE 15. Sample questions on a dynamic factor--window openings.

type, and measurement interval for previous studies on which preliminary estimates are based should parallel as closely as possible the conditions surrounding the contemplated monitoring program. Of course, in many instances there will be little or no information from previous studies. In these cases, some assumptions must be made or best judgment used as to expected levels and their variation. If logistic considerations permit, it may be prudent to apply a sequential sampling approach. Under this scheme, estimates obtained from the early portion of the study are used to refine the sample size for the latter part.

Once preliminary estimates of $\overline{X}$ and S have been made, the required sample size, n, can be approximated. The formula for sample size depends on whether the study has estimation or hypothesis-testing goals. A typical estimation goal is to estimate the average pollutant levels under prescribed conditions with a stated degree of precision. A typical hypothesis-testing goal is to compare pollutant levels from two differing sets of sampling conditions (e.g., two different types of structures) to test whether one of the conditions is associated with higher levels. The chances of arriving at incorrect conclusions on the basis of a statistical test are related to the chosen sample size.

In the case of estimation goals, a common statement of desired precision is as follows: "We wish to have a 95 percent confidence that the average level for the pollutant under consideration can be estimated within ±10 percent of its true value for the chosen sampling conditions." The formula for the sample size necessary to meet this objective is as follows:

$$n = \frac{t^2 S^2}{d^2} \tag{1}$$

where   t represents the number of standard deviations (approximately two) that account for the central 95 percent of the area under a normal curve

                S is the standard deviation for the variable to be estimated

                d is the margin of error (i.e., 10 percent of the true value).

The value for t in the above expression varies with the confidence level of choice. Given a confidence level, the approximate value for t can be found in an appendix of most statistical texts. As stated previously, best estimates of S and $\overline{X}$ are also required. The ratio $S/\overline{X}$ varies with sampling conditions but usually lies between 0.25 and 1.0 for CO, NO_2, and total suspended particulate matter (TSP). The ratio could be considerably larger for organic pollutants. The magnitude of the standard deviation is affected by the inherent variability of the environment as well as the imprecision of the measurement system.

If, for example, best estimates indicate that $S/\overline{X} = 0.5$, then $0.5\overline{X}$ can be substituted for S in the above expression. Because t = 2 and d = $0.1\overline{X}$ (i.e., 10 percent of the mean value), the required sample size is estimated as follows:

$$n = \frac{(2)^2(0.5\overline{X})^2}{(0.1\overline{X})^2} = \frac{4 \times 0.25\overline{X}^2}{0.01\overline{X}^2} = 100 \ . \tag{2}$$

Thus, for this hypothetical example, 100 air samples are required to achieve the desired precision.

When two sets of sampling conditions are to be statistically contrasted, a t-test is commonly used to test the null hypothesis that their concentration distributions arise from the same underlying distribution. Sample size estimates can often be obtained from the t-test specification, which has the following general form:

$$t = \frac{\overline{X}_2 - \overline{X}_1}{S\sqrt{\dfrac{1}{n_1} + \dfrac{1}{n_2}}} \tag{3}$$

where $\overline{X}_1$ and $\overline{X}_2$ are the mean concentrations for the two sets of sampling conditions

S is the standard deviation for the two sampling conditions (equal by assumption)

n_1 and n_2 are sample sizes for the two sampling conditions.

One important property of a statistical test is its power. (The power is the probability that a statistical test will detect a true difference in pollution levels for the two differing sampling conditions.) One minus the power is the probability (β) of making a type II error--concluding that two different sampling conditions have the same underlying concentration distributions, when they do not. The power of a statistical test increases as the size of the true, but unknown, difference between two sampling conditions increases. The other type of error is a type I error--concluding that two sampling conditions have different underlying distributions, when they do not. The probability of making a type I error is often denoted by α.

Type I and type II errors cannot be totally suppressed. For a fixed sample size, as α decreases, β increases, and vice versa. Thus, the sample size of choice and the α and β levels at which a statistical test is conducted are closely intertwined. Once two of these parameters are specified, the third is automatically determined. When providing study results, it is customary to report the level of significance (α level) at which the statistical test was conducted.

The α level for a statistical test should be specified before sampling is initiated. In choosing this level, the anticipated β error associated with the α level and the sample size of choice must be carefully considered. Depending on the situation, the consequences of type I errors, type II errors, or both, may be of genuine concern. In the above formula, both $\overline{X}_2$ and S can be expressed in relation to $\overline{X}_1$. Power curves found in statistical tests (e.g., Dixon and Massey 1969) can be used to relate the type I and type II errors associated with various sample sizes and assumed percentage differences between $\overline{X}_2$ and $\overline{X}_1$. The consequences of each type of error must be considered in choosing a sample size that will yield tolerable error levels.

The two example cases provided in Table 17 illustrate the considerations involved in choosing the appropriate sample size and level of significance for statistical testing. For both cases, it is assumed that $S/\overline{X}_1 = 0.5$. In the first case, a test is required to assess whether the two sets of sampling conditions yield average pollution levels that differ by 25 percent or more.

If 100 measurements are taken under each condition, then both α (0.05) and β (0.06) levels can be kept low. In the second case, a test to detect a smaller difference (10 percent or more) is required. In this case, 100 measurements for each condition do not appear to yield acceptable error probabilities. If 400 measurements

for each condition are taken, then α and β can be equalized at reasonably low
levels (0.10 and 0.11, respectively). If the test is performed at the 5 percent
level of significance (i.e., α = 0.05), then a β level of 0.19 can be anticipated.

TABLE 17. Estimated α and β Levels Associated with Selected Sample Sizes and Assumed Differences Between $\overline{X}_1$ and $\overline{X}_2$

Case 1. Test to detect whether $\overline{X}_1$ and $\overline{X}_2$ differ by 25 percent

Sample size		Error probabilities	
n_1	n_2	α	β
50	50	0.05	0.30
50	50	0.10	0.19
50	50	0.20	0.11
100	100	0.05	0.06
100	100	0.10	0.03
100	100	0.20	0.01

Case 2. Test to detect whether $\overline{X}_1$ and $\overline{X}_2$ differ by 10 percent

Sample size		Error probabilities	
n_1	n_2	α	β
100	100	0.05	0.70
100	100	0.10	0.59
100	100	0.20	0.44
400	400	0.05	0.19
400	400	0.10	0.11
400	400	0.20	0.07

In some cases, practical considerations may prevent attainment of the desired
sample size. For example, the above exercise might indicate that it is desirable
to obtain weekly samples from 200 houses during a single season. However, budgetary or logistic constraints may be such that only 15 monitoring devices can be
deployed per week over a period of 10 weeks, for a total of 150 weekly samples.
In this case, monitoring can still be conducted but in recognition that errors
associated with estimates or statistical tests will be higher than initially
planned. However, if the desired sample size diverges substantially from that
permitted by budgetary or logistic considerations, then the design may need to be
altered. An example alteration would be a reduction in the monitoring duration
from 7 days per house to 3 days per house to permit measurements in more houses,
recognizing that this approach will less precisely depict each house. As a
result, there could be greater house-to-house variations in monitoring results
and a lower potential for explaining these variations.

Sampling frame. A sampling frame is a list of the entire set of units from which
a sample is to be chosen for monitoring purposes. A sampling frame can be defined
in temporal terms, spatial terms, or both. For example, if a single house were

being monitored over an entire year, then the sampling frame would be the set of minutes, hours, or days comprising the year. On the other hand, if a cross-sectional study of houses was being performed at a specific time of year in a specific location, then the sampling frame would be the set of houses in that location at that time. Sampling frames can be restrictive, such as the set of days with temperatures averaging above 20 °C, the set of houses built since 1970, or those having a specific type of indoor appliance.

A sampling frame need not exhaust the entire set of units from which one wishes to sample. For example, a criss-cross or reverse telephone directory could be used as a sampling frame for a city or parts of the city. This frame might account for 80 to 90 percent or more of the existing housing units. As an alternative, certain neighborhoods within the city or blocks within those neighborhoods could be selected and the blocks could be visited to list all housing units. There is usually a tradeoff between the costs of developing a sampling frame and the completeness of the frame. Incomplete frames are less desirable if statistical inferences are to be made, especially if there is a bias in the nature of the incompleteness (e.g., houses without telephones being associated with lower socioeconomic classes).

A paper by Soczek et al. (1985) discusses various considerations related to sample size and sampling frame for a monitoring study involving a survey of residential NO_2 concentrations.

Step 6--Develop Protocols

Under this step, two types of protocols are developed in parallel--one for the selection of sampling units and one for the conduct of measurements. With the development of these protocols, the monitoring design is nearly complete; thus, costs can be estimated with greater precision than at the completion of step 3 (initial design).

Sample selection protocol. The sample selection protocol should specify how sampling units will be selected from the sampling frame to meet the desired sample size that was determined under step 5. Four types of samples distinguished by selection methods are as follows (Kish 1965):

- Fortuitous samples, such as those obtained by advertising for volunteers

- Judgmental samples, whereby experts exercise judgment in picking "representative" or "typical" sampling units

- Quota samples, whereby the sampling frame is continually tapped until quotas of sampling units of different types are found

- Probability samples, whereby sampling units are chosen at random from the sampling frame in such a way that their respective probabilities of selection are known.

Probability samples are desirable if statistical inferences are to be made. In some cases, probability sampling can be combined with judgment or quota sampling without seriously jeopardizing the ability to make inferences. For example, judgment could be used to select geographic areas and probability sampling could be used to select housing units within each area. As another example, probability sampling could be used to select housing units, but with the restriction that the number of housing units in different types of classes be equal. To accomplish this goal, the concept of quotas would be used. A useful device here is a

screening questionnaire, which is used to obtain a small set of information by
which housing units can be classified. Housing units are selected at random,
administered the screening questionnaire, and classified according to the questionnaire responses. This process continues until a targeted number of units is
found for each class. A specified number of units is then chosen at random from
the set of screened housing units in each class. The ultimate housing units
selected in this way are not likely to have equal probabilities of selection, but
their selection probabilities will be known.

It is very unlikely that all sampling units selected for monitoring will consent
to participate, unless the sample was obtained by advertising for volunteers. The
next step after selecting units is enrolling them in the monitoring study, which
requires contacts either in person, by telephone, or by mail. There are tradeoffs
in the methods of contact between resources expended and the response rate (that
is, the fraction of sampling units contacted that consent to participate in
monitoring). Response rates can often be improved by combining contact methods,
such as a letter followed by a telephone call or personal visit. Other incentives,
such as money, a gift, or a copy of the monitoring results, can help to improve
response rates. In general, designs that achieve higher response rates will be
less biased and will yield more generalizable results.

Measurement protocol. The measurement protocol should specify all measurement
devices and survey instruments to be used in the monitoring effort and should
describe the manner and sequence in which they will be deployed or administered.
Specific elements to be included in the protocol are as follows:

• Criteria for choosing indoor sampling zones and specific locations of
  monitoring devices or probes

• Criteria for locating outdoor devices or probes, if part of the monitoring
  effort

• Procedures for setting up and operating the monitoring devices

• Procedures for onsite recording of signals from monitoring devices

• Procedures for handling and transferring data records and exposed sample media

• Procedures for calibrating, making routine checks, and maintaining monitoring
  devices

• Procedures for analyzing exposed sample media

• Procedures for performing quality control checks and data quality assessments

• Forms for documenting the enactment and results of all procedures described
  above.

Operational considerations concerning probe placement and quality control and
quality assurance are provided in Chapter 6. The measurement protocol should
be reviewed by project personnel including the field staff to ensure that all
procedures are consistent and complete.

For survey instruments, the measurement protocol entails the specific formulation
of each item or question for which information is to be gathered. Where possible,
the range of responses to questions should be anticipated, so that similar answers
can be grouped to simplify later steps of coding responses. Questions should be
worded in a clear, straightforward manner so that the information is obtained
consistently from all types of respondents.

Step 7--Test Instruments, As Needed

In cases where the monitoring team has used all monitoring devices in previous research and is thoroughly familiar with their performance characteristics, this step may be bypassed. However, in some cases the devices will be used for the first time or will be applied to a new type of monitoring effort for which their performance characteristics are uncertain. In the latter case, some preliminary tests should be conducted before proceeding to the field. The purpose of this testing is to subject the devices to a range of concentrations and conditions similar to those expected in the field and to observe their performance. Precision checks can be made by colocating more than one device of the same type. Accuracy checks can be made by including as a reference point the most sophisticated devices available. Issues such as the extent of instrument drift over time or changes in instrument response due to temperature, humidity, or interferents can also be investigated.

Survey instruments can also be tested for proper flow and clarity by administering them to coworkers, friends, or other individuals. This limited testing prior to a pretest helps to confirm the range of responses that can be anticipated. But for a true test of clarity and respondent burden, the instruments must be administered to individuals who are totally unfamiliar with the research or investigation. By having different persons administer the instrument to the same individual, possible areas of interviewer bias can be identified. Similarly, by administering the same instrument to more than one person from the same building, questions that produce inconsistent responses can be identified.

Step 8--Conduct Training and Pretest

The intent of the pretest is to mimic as closely as possible the actual field monitoring effort, but on a much smaller scale. Although it may be tempting to skip this step in the interest of launching the monitoring effort as soon as possible, a pretest must be considered an unavoidable step. Furthermore, all elements of the sampling and measurement protocols should be applied in the pretest. Contacts with prospective respondents should be made in the same manner as planned for the actual study. The measurements with monitoring devices should be performed by technicians who will be conducting the actual field monitoring, and survey instruments should be administered by interviewers who will be used for the field effort. The pretest will not only help detect deficiencies in protocols or instruments but will also pinpoint areas of training needs; it will assist in a more realistic assessment of the costs and logistics associated with the full-scale monitoring effort.

A pretest must be considered as applied training after the needed conceptual or theoretical background is acquired. Proceeding directly to field monitoring without a pretest would be similar to taking a driving test with classroom training but no "behind-the-wheel" experience!

Step 9--Refine Protocols

If the monitoring design has been well thought through, protocol refinements based on the pretest would be generally small but still very useful. At this point, all protocols and instruments can be assembled or packaged in final form and any additional training efforts required for the monitoring can be initiated. However, in some cases the pretest may reveal deficiencies in instruments or unanticipated logistical requirements of such a nature that it is necessary to reassess the monitoring design. It may be necessary to return to step 4 or step 5 to address such problems.

The detailed design for monitoring resulting from step 9 is the culmination of all the preceding steps discussed in this section. The detailed design should include the refined protocols, the monitoring schedule, and final cost estimates.

In addition to field, laboratory, and office personnel, potential cost elements for the design are as follows:

- Instrumentation

- Laboratory analysis

- Quality control and quality assurance provisions

- Field travel and field office space

- Shipping and storage

- Communications (telephone, mail)

- Incentives for monitoring participants

- Data processing

- Forms and reports.

DESIGN CONSIDERATIONS FOR INVESTIGATING BUILDING-ASSOCIATED PROBLEMS

The investigation of building-associated problems often begins as a result of reported illnesses, symptoms, or complaints about air quality. In such cases, the immediate inclination is to conduct monitoring to identify the cause or the contaminants responsible for health-related or air quality problems. Yet experience shows that such a monitoring approach is seldom useful. The following points contribute to inconclusive investigations:

- Complaints by nature are subjective; hence, sorting out the useful information from possibly emotionally charged reports demands a systematic approach.

- Multiple etiologic factors, environmental factors, and even psychological factors may be responsible for complaints. Contaminants, if present, may be low level and difficult to identify and to relate to health symptoms.

Thus, seemingly straightforward investigations of building-associated problems become complex problems involving both people and their indoor environments. An emphasis on either problem area is unlikely to be productive.

The best approach in addressing such problems is to keenly observe and gather facts related to both the physical environment and people. In a practical sense, this is the approach used by detectives who carefully evaluate all factors that can provide a solution to the problem. Stolwijk (1984) recommends using a multi-disciplinary team of individuals with training and experience in ventilation engineering, measurement of contaminants, and health consequences of contaminants as well as in evaluation of social and psychological aspects of the building environment.

The relevant factors to be examined may include the following:

● Building ventilation and air exchange

● Indoor sources and other physical factors

● Complaints

● Complainants.

A systematic evaluation of observations and facts will narrow the many potential causes and help to pinpoint the problem.

The first stage of the evaluation should contain the following steps:

● Examine complaints for validity and consistency

● Evaluate the location of complaints to help determine the origin and source

● Collect information from persons with health complaints and compare descriptions with a control group

● Survey and evaluate ventilation systems including the location of exhausts with respect to intakes

● Survey unusual indoor sources or other physical factors and immediate outdoor environment that may cause the injection of contaminants into the indoor environment.

Although some of the professional associations are beginning to organize efforts to develop standardized methods and protocols for investigating building-associated problems, formal guidelines do not exist (American Conference of Governmental Industrial Hygienists (ACGIH) 1985). Ferrand et al. (1985) were among the first to attempt to develop a preliminary protocol to assist in such investigations; their approach includes a background questionnaire and forms for onsite visit and environmental monitoring. The questions in their questionnaire can be regrouped to address the following fundamental points:

● Who were affected and how? What type of individuals (i.e., supervisors, clerical, visitors) were affected? What were their symptoms? At what locations do the affected people work? Is there a group of individuals having similar characteristics as the affected group in the same building and served by the same heating, ventilating, and air conditioning (HVAC) system but who were not affected?

● When did the problem occur? How long has the problem been in existence? Does it occur periodically (that is, at a certain time of day, on a certain day of the week, or during a certain season)? Does it coincide with changes or repairs that may produce contaminants?

● What are the potential sources of contaminants in the building? Were any new equipment or materials introduced or was any existing equipment modified? Were any new furniture, accessories, or decorative materials or coatings introduced? What are the cleaners, detergents, disinfectants, and insecticides used in the building and what is the frequency of their use?

- What type of HVAC system does the building have? What is the percentage of the recirculation air? Can windows be opened? Are there other parts of the building that share the same ventilation as the affected area? Were any changes made in the HVAC system around the time of the incident?

- What are the potential sources of contaminants outside the building? Are there any sources near the fresh-air intake for the building? Are there any underground or other parking facilities that connect with the building?

One of the important areas that has not been addressed by Ferrand et al. (1985) is the potential for biological contaminants resulting from an improperly functioning HVAC system or from flooding. As a matter of fact, a number of investigations carried out by the National Institute for Occupational Safety and Health have shown microbiological contaminants to be the cause of respiratory and sensory irritation and a number of diseases. Morey et al. (1984) have found that several buildings were characterized by a history of repeated flooding and all contained mechanical systems with pools of stagnant water and microbial slimes.

Answers to the five fundamental questions listed above will narrow the scope of the investigation. In general, no extensive measurements should be conducted before such questions are addressed. One possible exception is measurements of ventilation rates and ventilation-related parameters such as CO_2 for evaluation of the ventilation system (Kreiss and Hodgson 1984).

Once the possibilities are narrowed, the investigator can proceed in one of two ways: (1) use the collected information to alter possible conditions related to the problem through a trial-and-error approach or (2) continue the investigation and include the use of monitoring programs to pinpoint the causes. For monitoring, although the urgency of responding to the problem may preclude explicit consideration of the nine steps outlined earlier in this chapter, effort must be made to follow the general principles and to develop the preliminary research plan.

As pointed out earlier, some professional associations (in particular, ASTM and ACGIH) are organizing committees to standardize protocols for investigating building-associated illnesses. These organizations can be contacted for further information (see Chapter 7).

Some preventive measures that may be effective in reducing building-associated microbial contamination and building-associated hypersensitivity pneumonitis illnesses include the following:

- Preventing moisture buildup in occupied space and HVAC system components

- Removing stagnant water and slimes from building mechanical systems

- Using steam as a moisture source in humidifiers

- Eliminating the use of water sprays as components of office building HVAC systems

- Keeping relative humidity below 70 percent

- Using filters with a 50- to 70-percent rated efficiency

- Discarding microbially damaged office furnishings

- Initiating a fastidious maintenance program for HVAC system air-handling units and fan coil units.

REFERENCES

American Conference of Governmental Industrial Hygienists (ACGIH), Report of ACGIH Ad Hoc Committee on Bioaerosols, ACGIH, Cincinnati, 1985.

Dixon, W.F., and Massey, J.J. Jr., Introduction to Statistical Analysis, McGraw-Hill Book Company, New York, 1969.

Ferrand, E.F., Nichik, F., Padnos, N., and Beck, E.M., Urban Indoor Air Investigative Protocol, Proceedings of the 78th Annual Meeting of the Air Pollution Control Association, Pittsburgh, paper no. 85-46.8, 1985.

Fortmann, R.C., and Nagda, N.L., Work Plans for Phase II of EPRI Project 2034-1, Energy Use, Infiltration, and Indoor Air Quality in Tight, Well-Insulated Residences, Report Number ER-1585, GEOMET Technologies, Inc., Germantown, Maryland, 1985.

Kish, L., Survey Sampling, John Wiley & Sons, Inc., New York, 1965.

Kreiss, K., and Hodgson, M.J., Building-Associated Epidemics, in Indoor Air Quality, ed. P.J. Walsh, C.S. Dudney, and E.D. Copenhaven, pp. 87-106, CRC Press, Boca Raton, Florida, 1984.

Morey, P.R., Hodgson, M.J., Sorenson, W.G., Kullman, G.J., Rhodes, W.W., and Visvesvara, G.S., Environmental Studies in Moldy Office Buildings: Biological Agents, Sources, and Preventive Measures, Ann. Am. Conf. Gov. Ind. Hyg., vol. 10, pp. 21-35, 1984.

Nagda, N.L., Koontz, M.D., Rector, H.E., Harrje, D., Lannus, A., Patterson, R., and Purcell, G., Study Design to Relate Residential Energy Use, Air Infiltration and Indoor Air Quality, Proceedings of the 76th Annual Meeting of the Air Pollution Control Association, Pittsburgh, paper no. 83-29.3, 1983.

Nagda, N.L., Koontz, M.D., and Rector, H.E., A Field Monitoring Study of Homes with Unvented Gas Space Heaters: Preliminary Research Plan, Report Number ER-1489, GEOMET Technologies, Inc., Germantown, Maryland, 1984.

Saltzman, B.E., Significance of Sampling Time in Air Monitoring, J. Air Pollut. Control Assoc., vol. 20, no. 10, pp. 660-665, 1970.

Soczek, M.L., Ryan, P.B., Spengler, J.D., Fowler, F.J., and Billick, I.H., A Survey Methodology for Characterization of Residential NO_2 Concentrations, Proceedings of the 78th Annual Meeting of the Air Pollution Control Association, Pittsburgh, paper no. 85-31.2, 1985.

Stolwijk, J.A.J., The "Sick Building" Syndrome, Proceedings of the 3rd International Conference on Indoor Air Quality and Climate, Stockholm, vol. 1, p. 23, 1984.

Chapter 6
PRACTICAL CONSIDERATIONS

Successful implementation of a monitoring design often hinges on factors that are incompletely evaluated during the design stage. This chapter addresses some potential pitfalls and pratfalls under the broad themes of representative sampling and quality assurance.

REPRESENTATIVE SAMPLING

Indoor and outdoor probe locations should be selected to represent the environment in a manner that is consistent with objectives. For simplicity, two types of monitoring objectives are considered:

1. Characterization--to represent peak or average concentrations in the context of the indoor mass balance, and

2. Exposure--to represent peak or average concentrations experienced by an individual occupying the airspace.

For characterization studies, monitoring takes place at one or more fixed locations indoors that are selected based on mass balance criteria. Exposure studies can also use fixed monitors, but indoor probe locations are selected based on occupant activities. If monitors are sufficiently miniaturized to be comfortably worn or carried about, exposure can be assessed by having the instrument "ride" with the monitoring subject.

Selecting a probe location for stationary monitoring is a two-step procedure. The first step is to select a zone; the zone may be a general indoor airspace such as the main floor or the basement or may be a specific room such as a kitchen or living room. The second step is to select a specific location in that zone that will serve to represent, with respect to objectives, the entire zone.

In many cases, the first step of selecting a zone will be explicitly specified in a study objective (that is, measurements will be taken in rooms containing sources of type X). In studies involving occupied structures, parallel interviews or, for consistency when sampling a number of structures, carefully prepared survey instruments are invaluable in identifying structural characteristics. These characteristics include age and building materials, contents such as appliances and furnishings, occupant practices such as use of appliances, and potential interferents. Moreover, the habits of occupants with respect to indoor locations are of particular concern for exposure studies because zone selection is strictly tied to occupancy patterns.

There are no generalized guidelines available for selecting zones if zones are not specified in study objectives. Normally, the number of zones to be monitored is restricted because of resource considerations.

Thus, in the absence of guidelines and with a desire to limit the number of zones to be monitored, the anticipated mass balance for the pollutant in question must be qualitatively conceptualized. The locations of sources, the effective volume available for dilution of source emissions, and air exchange are some of the factors that should be considered. Additionally, considerations such as thermal buoyancy and the potential effect, if any, of the air-handling system on circulation are important. For example, warm air emitted from unvented combustion appliances tends to rise nearly vertically until a barrier to the rise is encountered; thus, in a two-story house, if the appliance is on the lower floor, the upper floor will receive combustion gases through an open stairwell almost as rapidly as these gases mix throughout the lower floor (Nagda et al. 1985).

Adaptation of the tracer-gas techniques introduced in Chapter 4 can provide useful insights for zone selection. Simultaneous measurements of tracer-gas decay at a number of indoor locations have been used to evaluate zone-to-zone differences in air exchange rates and to infer interzonal flows (Maldonado and Woods 1983). Such an approach can identify zones of relatively higher or lower air exchange rates.

Multiple tracers, however, appear to be the only quantitative approach to addressing zone selection questions that involve air exchange among indoor zones. Given a large number of structures to be monitored, this approach can be quite resource intensive, but the resources expended can be minimized by limiting applications to only a few typical structures. These structures should be representative of those intended to be monitored in terms of structural characteristics, contents of the structure, and occupant practices.

Once indoor zones have been identified, probe locations may be selected. In selecting probe locations, the following areas should be avoided:

● Exterior walls and corners

● Areas that receive direct sunlight

● Areas where there are noticeable drafts

● Areas directly influenced by supply or return ducts

● Probe heights below 1 m or above 2 m unless vertical gradients are being measured

● Well-trafficked spots

● Areas that receive direct impact from indoor sources.

Many indoor air quality studies require simultaneous measurements of outdoor concentrations. Probe siting criteria have been established for monitoring outdoor concentrations (EPA 1979). However, such criteria may need to be selectively compromised because indoor air quality studies focus on the nearby outdoor air that infiltrates into the structure, whereas most ambient outdoor monitoring is concerned with the representation of a larger region.

Instrumentation factors play a large role in selecting probe locations, too. Monitors should be configured to interfere with the indoor environment as little

as possible. For unoccupied structures, primary considerations include flow
rates (to avoid having the sampling system act as an air cleaner or as a local
exhaust), and the evolution of heat. For occupied structures--particularly
occupied residences--available space also becomes an important issue.

In some studies that have examined indoor/outdoor air quality relationships,
mobile laboratories have been used that were simply parked next to the structure
(Moschandreas et al. 1981). Sample lines were strung from the mobile laboratory
into the building to allow monitoring of one or more indoor zones from a single
location.

This approach has allowed flexible operations, sophisticated instrumentation,
centralized data collection, and onsite workspace. However, this approach is
not universally applicable; many neighborhoods simply do not provide sufficient
space to park a mobile laboratory.

The mobile laboratory approach also consumes a relatively large amount of elec-
tricity for instruments, lighting, and environmental control--so much power that
a substantial amount of the scheduling may be determined by local power companies
providing special power drops. Most mobile laboratory systems cannot be run from
an extension cord. Nonetheless, instrument packages of this type are of great
use if resources and study conditions permit.

At the other end of the spectrum, many newer analyzers are sufficiently miniaturized to be packaged for placement indoors to operate from battery power or to
operate from household electric supply without interfering with normal occupancy.
However, these systems usually require some sort of repackaging to facilitate use
in the field. For systems featuring multiple analyzers and sophisticated data
recording, some type of box is useful for easing transport and providing security.
However, it is very easy at the design stage to slip into a "boat in the basement" problem when configuring systems. These are just some of the questions
that must be answered early:

● How many people will it take to lug the measurement package about?

● What is the size of the smallest doorway through which it must be carried
  (including vehicles used to transport the package from place to place)?

● Can a toddler pull it over?

● Will the size of the package interfere with normal use of the zone by
  occupants?

● Will the package emit noise or odors that may be considered offensive to
  occupants?

If the system is to be run from wall current, electric power is important for
two reasons. The first is heat: transformers, pumps, etc., will generate heat
during operation. If repackaging confines natural ventilation around instruments, the casing should provide for compensatory air movement with small fans
or strategically placed louvers. However, if sampling inlets are very close to
the cabinetry, sampling may be biased.

The second aspect of electric power is the system amperage and grounding requirements. If monitoring is to take place in occupied structures, available circuits
will be at a premium. A blown fuse or tripped breaker leads to lost data and
guilt-ridden, if not infuriated, occupants. There are many structures that still
have two-prong outlets: a "cheater plug" does not necessarily ensure a grounded
connection. Inexpensive test devices are available to verify ground connections.

From an installation standpoint, passive collectors offer the easiest route. They are generally small enough to be placed out of the way by means of string and a thumbtack. Many such commercial devices come equipped with spring-loaded clips or a moderate length of ribbon to simplify attachment.

The use of passive collectors is not without difficulties, however. Precision (the degree of mutual agreement from repeated measurements of the same condition) for some passive collectors can be as low as 25 percent--especially when low sample mass is involved. Distinguishing differences (and gradients) versus similarities requires a good understanding of the precision and accuracy as well as the statistics involved. Passive samples should be collected in duplicate or even triplicate as often as possible to allow evaluation of precision on a routine basis.

QUALITY ASSURANCE

There are two kinds of measurement variability--that which is attributable to occurrences in the measurement scene and that which is attributable to performance of the measurement system. Measurement variability of the first type forms a principal link between data and objective. For example, measured indoor concentrations can increase when indoor sources are present. Mass balance relationships can be used to infer emission rates. Measurement variability of the second type, if unrecognized or poorly estimated, affects interpretation or can even lead to erroneous interpretations. For example, a failing instrument component or inadequate calibration can produce data signals that appear reasonable but are not related to actual concentrations.

Systematic treatment of the question of data reliability is addressed through quality control (QC) and quality assurance (QA) procedures. QC enters the measurement process in the form of calibrations and checks to routinely measure the reliability of data. For completeness, QA operations test the reliability of the QC process.

Provisions for QA and QC are implemented through a QA plan. The plan specifies in detail the manner in which a particular project or continuing operation will achieve predetermined goals of data quality. Often, client policy or regulation will require a QA plan. Even if a QA plan is not formally required, however, the process of developing a plan forces an investigator to review every operational aspect in an orderly manner. This review strengthens the approach to the monitoring problem as well as the resultant data.

The EPA (1983) has published guidelines for preparing QA plans. These guidelines define the principal elements of QA plans for general environmental measurements and are applicable to indoor air quality monitoring.

QA plans revolve around five basic issues: (1) QA objectives, (2) custody procedures, (3) internal QC checks, (4) performance and systems audits, and (5) corrective actions. Each of these is briefly discussed below. More thorough explanations may be found in references cited in the EPA guidelines and in the example QA plan listed at the end of this section.

Quality Assurance Objectives

QA objectives define the level of instrument performance variability that is acceptable. They are implicitly related to monitoring objectives and to the remainder of the design because they define the limits of interpretation. For each measurement variable, QA objectives should be defined in terms of the following:

- Accuracy--the degree of agreement of a measurement with an accepted reference or true value

- Precision--a measure of mutual agreement among individual measurements under prescribed conditions

- Completeness--compared with expectations, a measure in percent of the amount of valid data recovered

- Representativeness--an expression of the degree to which data represent key characteristics or conditions

- Comparability--an expression that defines the degree of confidence with which one data set can be compared with another.

QA objectives may be drawn from experience as well as from standard methodologies. Pretests involving part or all of the measuring systems are often useful for determining QA objectives for untested or modified techniques. The QA plan should also address routine procedures to assess precision, accuracy, and completeness of the accumulating data for each major measurement parameter. The results of the assessment must be continually tested against QA objectives.

Custody Procedures

Custody procedures provide safeguards against data loss and are particularly useful in determining sources of contamination or other adverse factors that might jeopardize the quality of data. In some cases, strict chain-of-custody procedures are required if the data are to be treated as legal evidence. Even if strict custody procedures are not required, it is nonetheless useful to clearly document the paths taken by all relocatable elements and to specify handling procedures at each custody point. A relocatable element is any item that affects the final data product, such as sample media, primary data records, and reduced data records. Where appropriate, each relocatable element receives a unique identity, including serial number, date, time, and location. A log should document the movement of the element among various points of custody--technical personnel, files, and storage.

Internal Quality Control Checks

Internal QC checks consist of periodic testing of equipment performance and assessment of procedures. For approaches relying on sample collection and laboratory analysis, the following types of checks should be considered: (1) replicates, (2) spiked samples, (3) split samples, (4) blanks, and (5) reagent checks. For approaches relying on continuous analyzers, routine multipoint calibrations, zero-span checks, and colocated monitoring are warranted. The QC checks should evaluate all procedures and equipment under conditions of ordinary use.

Performance and Systems Audits

Performance audits should be conducted periodically to determine the accuracy of the total measurement system and its individual components. Most aspects of a performance audit are similar to those of the internal QC checks except that performance is verified through standards, devices, and personnel that are independent of the routine project organization and equipment.

Systems audits consist of a qualitative evaluation of the facilities, equipment, training, procedures, recordkeeping, data validation, and reporting aspects of the total monitoring approach. This evaluation assesses the capability to perform within QA objectives.

Both performance and systems audits should precede initial data collection. Thereafter, the frequency of audits is dictated by policy, objectives, and resources.

Corrective Action

Corrective actions are the systematic responses to errors, malfunctions, and other deficiencies. Corrective actions may result from internal QC checks as well as from performance or systems audits. The QA plan should stipulate procedures to be followed in correcting a deficiency. Regardless of the size of the deficiency or eventual need for corrective action, the sequence for corrective action is composed of three phases:

1. Analysis--to determine potential causes, to assess the extent of negative impact on accumulated data, and to identify reasonable corrective actions,

2. Adjustment--to transmit corrective steps to cognizant personnel and to adjust, relabel, or discard affected data, and

3. Report--to document the entire corrective operation in the permanent records.

Example of a Quality Assurance Plan

It is useful to review QA plans prepared for other studies. A plan prepared for the EPA (Research Triangle Institute 1981) is a model example of an operational QA project plan. It follows the EPA guidelines and specifications closely. The plan is directed toward field sampling and laboratory analysis; but many, if not all, the items in the QA plan apply to direct field and/or laboratory measurements as well. For continuing operations, a QA program plan (see GEOMET 1985, for example) is recommended to provide for consistency across projects.

REFERENCES

EPA. See U.S. Environmental Protection Agency.

GEOMET Technologies, Inc., "Quality Assurance Plan for Indoor Environment Program," GEOMET Report No. ES-1528, GEOMET Technologies, Inc., Germantown, Maryland, 1985.

Maldonado, E.A.B., and Woods, J.E., "A Method to Select Locations for Indoor Air Quality Sampling," Build. Environ., vol. 18, no. 4, pp. 171-180, 1983.

Moschandreas, D.J., Zabransky, J., and Pelton, D.J., "Comparison of Indoor and Outdoor Air Quality," Report No. EA-1733, research project 1309, Electric Power Research Institute, Palo Alto, California, 1981.

Nagda, N.L., Koontz, M.D., and Billick, I.H., "Effect of Unvented Combustion Appliances on Air Exchange Among Indoor Spaces" in Ventilation Strategies and Measurement Techniques: Proceedings (Document AIC-PROC-6-85), Air Infiltration Centre, Bracknell, Berkshire, Great Britain, 1985.

Research Triangle Institute, Total Exposure Assessment Methodology (TEAM) Study. Phase II/Part III: Quality Assurance Project Plan prepared under Contract No. 68-02-3679 for U.S. Environmental Protection Agency, Research Triangle Park, North Carolina, 1981.

U.S. Environmental Protection Agency, Ambient Air Quality Monitoring, Data Reporting and Surveillance Provision, 40 CFR 58, Appendix E: Probe Siting Criteria for Ambient Air Quality, Federal Register (44)92:27592-27597, 1979.

U.S. Environmental Protection Agency, Interim Guidelines and Specifications for Preparing Quality Assurance Project Plans, EPA-600/4-83-004, NTIS PB83-170514, Office of Exploratory Research, Washington, D.C., 1983.

Chapter 7
OTHER INFORMATION SOURCES

In addition to the guidelines offered in this book, there are numerous other
sources of information that describe pollutant and building behavior; methods
and devices for pollutant measurement, modeling, and control; and results of
past research. These information sources can be classified as scientific
societies and organizations; books, publications, or proceedings concerning
the topic of indoor air quality; and books or publications concerning measurements. Such sources are briefly described in this chapter to point the reader
in appropriate directions for supplemental information.

SCIENTIFIC SOCIETIES AND ORGANIZATIONS

1. American Society for Testing and Materials (ASTM)
    1916 Race Street
    Philadelphia, Pennsylvania 19103
    (213) 299-5400

ASTM is a scientific and technical organization that was formed in 1898 for the
development of standards on characteristics and performance of materials, products, systems, and services. Standard test methods, definitions, recommended
practices, classifications, and specifications are developed in ASTM through a
consensus-forming approach. In 1985 an ASTM Subcommittee on Indoor Air (D22.05)
was organized to promote knowledge and formulation of standard terminology, test
methods, practices, and guidelines for the sampling and analysis of indoor air.
Several task groups have been formed within Subcommittee D22.05 to address
different groups of pollutants and topic areas such as designs and protocols and
source characterization. Some other ASTM subcommittees address topics that are
related to indoor air quality; for example, Subcommittee E6.41 on infiltration
performance has promulgated standard methods for measuring building tightness
(blower door) and air infiltration (tracer gas), as discussed in Chapter 4. ASTM
periodically releases special technical publications (STPs) and standards books.

2. Air Pollution Control Association (APCA)
    P.O. Box 2861
    Pittsburgh, Pennsylvania 15230
    (412) 621-1090

APCA was founded in 1907 to promote clean air, provide leadership in the field of
air pollution control, and promote a sense of environmental responsibility. A
committee on indoor air quality (TT-7) has been in existence since 1977; its
purpose is to promote greater interest in and understanding of indoor air quality

and to help determine the importance of indoor air quality as part of the total human exposure to air contaminants. The committee has sponsored two to four sessions at annual APCA meetings. In 1985, indoor air quality became one of the major themes of APCA, and emphasis was significantly increased. Committee work has been divided into seven areas, including health effects, monitoring and modeling, and public policy. The Journal of the Air Pollution Control Association is a monthly APCA publication.

3. American Society of Heating, Refrigerating, and Air Conditioning Engineers
   (ASHRAE)
   1791 Tullie Circle, NE.
   Atlanta, Georgia 30329
   (404) 636-8400

ASHRAE was formed in 1894 to advance the fields of heating, refrigerating, air conditioning, and related human factors for the benefit of the general public. It provides forums for the discussion of technological developments, sponsors research, and develops voluntary consensus standards for uniform test methods, terminology, and recommended practice. One of these consensus standards, ASHRAE 62-1981, has established minimum ventilation rates for acceptable indoor air quality. Other standards have been developed in such areas as energy conservation and comfort conditions. ASHRAE Transactions, published annually, contains research and application papers presented at annual meetings. The ASHRAE Handbook, organized into four volumes--Fundamentals, Equipment, Systems, and Applications-- forms a comprehensive information resource. ASHRAE Journal is a monthly publication of the society.

4. American Industrial Hygiene Association (AIHA)
   475 Wolf Ledges Parkway
   Akron, Ohio 44311-1087
   (216) 762-7294

AIHA was founded in 1939 to promote the study and control of environmental factors affecting the health and well-being of industrial workers. Although AIHA is primarily concerned with industrial settings, many methods and concepts from this area are transferable to commercial or residential environments. The association publishes a monthly American Industrial Hygiene Association Journal and maintains a list of accredited laboratories. The association's laboratory accreditation program is based on laboratory participation in the National Institute for Occupational Safety and Health (NIOSH) proficiency testing program (PAT), which includes bimonthly samples for lead, cadmium, zinc, asbestos, silica, and organic solvents. Among AIHA's technical committees is one on indoor environmental quality.

5. American Conference of Governmental Industrial Hygienists (ACGIH)
   6500 Glenway Avenue
   Building D-5
   Cincinnati, Ohio 45211
   (513) 661-7881

ACGIH was organized in 1938 to develop administrative and technical aspects of worker health protection. Membership is limited to professional personnel in governmental agencies or educational institutions engaged in occupational safety and health programs (the conference is not an official Government agency). In addition to the transactions of annual meetings, the Conference publishes a number of manuals, guides, and reports, such as Air Sampling Instruments for Evaluation of Atmospheric Contaminants. ACGIH has formed a committee on bioaerosols to develop protocols for industrial environments.

6. Air Infiltration Centre (AIC)
    Old Bracknell Lane West
    Bracknell
    Berkshire, Great Britain, RG124AH
    International Telephone: 44 344 53123

An agreement formulated among a number of industrialized nations in 1974 established the International Energy Agency (IEA) as an autonomous body within the
Organisation for Economic Cooperation and Development. IEA inaugurated the AIC
with the aims of promoting the understanding of fundamental air infiltration processes and advancing the application of energy saving measures while ensuring the
preservation of safe and healthy indoor environments. AIC produces a quarterly
newsletter entitled Air Infiltration Review and releases a Technical Note series
of publications concerning infiltration and related topics. An example is
Technical Note 12, entitled "1983 Survey of Current Research Into Air Infiltration
and Related Air Quality Problems in Buildings." The AIC library includes a
bibliographic data base, called AIRBASE, that includes abstracts of papers on air
infiltration and ventilation aspects of indoor air quality.

BOOKS/PUBLICATIONS ON INDOOR AIR QUALITY

1. Meyer, B., Indoor Air Quality, Addison-Wesley Publishing Co., Inc., Reading,
    Massachusetts, 1983.

The intent of this book is to provide the reader with an understanding of some of
the many factors that cause indoor air pollution. The book covers a considerable
number of topics related to air quality; consequently, none are treated in depth.
Topics include a historical perspective, comfort and climate considerations,
building design and materials, methods for air monitoring and analysis, indoor
concentrations measured in selected studies, tools for indoor air quality
control, and regulatory provisions that directly or indirectly affect indoor
air quality. References in the text cover the period through March 1982.
The book contains a subject and author index.

2. Turiel, I., Indoor Air Quality and Human Health, Stanford University Press,
    Stanford, California, 1985.

Aimed at a wide audience, this book summarizes the health risks associated with
indoor air quality and discusses control techniques. Following an introduction
to basic concepts of indoor air quality and health effects, separate chapters are
devoted to formaldehyde and other chemical contaminants, radon, particulate
matter, combustion products, and involuntary smoking, leading to discussion of
the role of energy efficiency and a survey of pollutant control techniques.
Problems associated with office buildings are also presented. Legal and
regulatory issues are discussed in the final chapter. This book also features an
informative glossary. The references and suggested reading list include works
as recent as 1983.

3. Wadden, R.A., and Scheff, P.A., Indoor Air Pollution: Characterization,
    Prediction, and Control, John Wiley & Sons, New York, New York, 1983.

This book reviews problems that can occur in indoor environments in domestic and
public buildings and supplies tools for their solution. Included are discussions
of health effects criteria, outdoor and indoor contributions to indoor air quality,
and measurement considerations. The mass-balance equation is described as a basis
for formulating predictive indoor air quality models, and a few examples are
worked out to illustrate applications of the models. Indoor air quality standards
and control methods are also discussed. References through 1982 are included.

4. Walsh, P.J., Dudney, C.S., and Copenhaver, E.D. (eds.), Indoor Air Quality, CRC Press, Inc., Boca Raton, Florida, 1984.

This book is a wide-ranging review of indoor air quality. Each chapter, written by one or more leading researchers in the field, summarizes a particular topic. The text is divided into three sections. The first section, Generic Aspects, introduces theoretical and technical fundamentals and includes chapters on measurement techniques and health risk assessment. The second section, Phenomenological Aspects, contains chapters covering indoor air quality in typical residences and in energy-efficient residences as well as an in-depth review of building-associated epidemics. The final section summarizes pollutant-specific aspects for formaldehyde, radon, tobacco smoke, and allergens and pathogens. References date from 1982 and earlier. Although some topics such as modeling concepts and control strategies are treated rather briefly, the pollutant-specific discussions--particularly for bioaerosols and building-associated epidemics--are extensive.

5. U.S. Department of Energy, Indoor Air Quality Environmental Information Handbook: Combustion Sources (prepared by Mueller Associates, Inc., SYSCON Corporation and Brookhaven National Laboratory), Publication No. DOE/EV/10450-1, Washington, D.C., 1985.

This publication summarizes the current state of knowledge regarding combustion sources of indoor air pollution. Topics that are discussed include health effects, emission characteristics of combustion sources, factors affecting pollutant concentrations, indoor monitoring techniques, and results of selected monitoring programs. A review of models, controls, and standards applicable to indoor air quality are also presented. The final section of the publication offers a less technical perspective on these issues that is targeted toward homeowners. References through July 1984 are included.

6. Electric Power Research Institute, Manual on Indoor Air Quality (prepared by Lawrence Berkeley Laboratory), Publication No. EPRI EM-3469, Palo Alto, California, 1984.

This manual was prepared to assist electric utilities in helping homeowners, builders, and new home buyers to understand a broad range of issues related to indoor air quality. Topics covered include factors affecting air exchange and indoor air quality, sources and measured concentrations of key indoor pollutants, health effects and standards, and techniques for monitoring, modeling, and controlling indoor air quality. References through 1983 are included.

7. Sandia National Laboratories, Indoor Air Quality Handbook for Designers, Builders, and Users of Energy-Efficient Residences, Sandia 82-1773, Albuquerque, New Mexico, 1982.

This handbook was produced to assist designers, builders, and users of energy-efficient residences in achieving the apparently conflicting goals of energy efficiency and good indoor air quality. In an easy-to-understand style, the handbook covers a variety of topics such as influences of structure, contents, and air exchange on indoor air quality, characteristics of indoor air contaminants including health effects, control of contaminants, and legal aspects. References date from 1981 and earlier. The manual also provides a glossary and subject index.

8. National Research Council, Committee on Indoor Pollutants, Indoor Pollutants, National Academy Press, Washington, D.C., 1981.

The National Research Council (NRC) report, containing over 500 pages, is a comprehensive review and appraisal of indoor air quality literature. This report

includes chapters on sources and characterization of indoor pollutants, factors
that influence exposure, monitoring and modeling, health effects, welfare effects,
and control of indoor pollution. Recommendations for further research are presented, and an extensive list of references follows each chapter. The NRC report
lacks author or subject indexes, but is an excellent review of literature
through 1980.

CONFERENCE PROCEEDINGS AND SPECIAL JOURNAL ISSUES RELATED TO INDOOR AIR QUALITY

1. Measured Air Leakage Performance of Buildings (ed., H.R. Trechsel), American
   Society for Testing and Materials, ASTM STP No. 904, 1986.

2. Proceedings: Indoor Air Quality Seminar--Implications for Electric Utility
   Conservation Programs, EPRI Publication No. EA/EM-3824 (prepared by William
   I. Whidden and Associates and Hart, McMurphy and Parks, Inc.), 1985.

3. Indoor Air: Proceedings of the 3rd International Conference on Indoor Air
   Quality and Climate, Swedish Council for Building Research, Stockholm, Sweden,
   1984.

   Volume 1: Recent Advances in the Health Sciences and Technology
   Volume 2: Radon, Passive Smoking, Particulates, and Housing Epidemiology
   Volume 3: Sensory and Hyperreactivity Reactions to Sick Buildings
   Volume 4: Chemical Characterization and Personal Exposure
   Volume 5: Buildings, Ventilation, and Thermal Climate

4. Environment International, vol. 8, nos. 1-6, special issue: "Indoor Air
   Pollution." Compendium of papers from Pergammon Press, Elmsford, New York,
   1982.

5. Health Physics, vol. 45, no. 2, "Special Issue on Indoor Radon." Pergammon
   Press, Elmsford, New York, 1983.

INFORMATION SOURCES FOR MONITORING DEVICES

References and information sources for instruments and methods include scientific literature describing fundamental technologies, catalogs and directories
describing products, and manufacturers' literature on individual products.
Useful sources of information in the scientific literature include the following:

1. American Society of Heating, Refrigerating and Air Conditioning Engineers.
   ASHRAE Handbook - 1985 Fundamentals, American Society of Heating, Refrigerating and Air Conditioning Engineers, Atlanta, Georgia, 1985.

2. Finkelstein, P.L., Mazzarella, D.A., King, W.J., and White, J.H., Quality
   Assurance Handbook for Air Pollution Measurement Systems. Volume IV, Meteo-
   rological Instruments, EPA-600/4-82-060, U.S. Environmental Protection
   Agency, Environmental Measurements System Laboratory, Research Triangle Park,
   North Carolina, 1983.

3. Katz, M. (ed.), Methods of Air Sampling and Analysis, 2d ed., American
   Public Health Association, Washington, D.C., 1977.

4. Linch, A.L., Evaluation of Ambient Air Quality by Personal Monitoring, vol. I:
   "Gases and Vapors," and vol. II: "Aerosols, Monitor Pumps, Calibration, and
   Quality Control," CRC Press, Inc., Boca Raton, Florida, 1981.

5. Lioy, P.J., and Lioy, M.J.Y. (eds.), Air Sampling Instruments for Evaluation
    of Atmospheric Contaminants, 6th ed., American Conference of Governmental
    Industrial Hygienists, Cincinnati, Ohio, 1983.

6. Stern, A.C. (ed.), Air Pollution, 3d ed., vol. III, "Measuring, Monitoring
    and Surveillance of Air Pollution," Academic Press, New York, 1977.

7. Taylor, J.K., and Stanley, T.W., Quality Assurance for Environmental
    Measurements, ASTM STP 867, American Society for Testing and Materials,
    Philadelphia, Pennsylvania, 1985.

Especially useful are the professional journals such as the Journal of the Air
Pollution Control Association, the American Industrial Hygiene Association
Journal, and Analytical Chemistry that periodically offer reviews and information
on recent developments. These journals often refer to additional literature on
instruments and methods.

Examples of consolidated catalogs include Pollution Equipment News, published
seven times a year, and Industrial Hygiene News, published six times a year.
Reinbach Publication of Pittsburgh, Pennsylvania, circulates both without charge
to qualified subscribers. Each catalog updates a number of products and an
annual buyer's guide cross-references manufacturers by their products. Some
professional societies also publish annual directories listing instrument manufacturers by their products. Examples include the Directory and Resource Book
from the Air Pollution Control Association and the Guide to Scientific Instruments from the American Association for the Advancement of Science.

Finally, many instrument manufacturers publish technical notes covering instrument operation, special applications, and other information. Many references
cited for individual instrument summaries in Appendix A include such manufacturers' notes.

Appendix A
COMMERCIALLY AVAILABLE INSTRUMENTATION

This appendix reviews some of the commercially available instruments that are suitable for monitoring pollutants indoors and, in some cases, for personal exposure monitoring. Considering all combinations of mobility (personal/portable/stationary), power requirements (active/passive), and output characteristics (analyzer/collector), up to 12 instrument types are theoretically possible for any pollutant.

There is no pollutant for which all 12 types of instruments are available. In most cases, however, there are alternatives among personal, portable, and stationary instruments. Moreover, technological advances and commercialization of research products are continually resulting in new devices.

Individual summaries are alphabetically organized by pollutant (asbestos and other fibrous aerosols, biological aerosols, carbon monoxide (CO), formaldehyde (HCHO), inhalable particulate matter (IP), nitrogen dioxide (NO_2), ozone (O_3), radon (Rn), and sulfur dioxide (SO_2)) and by manufacturer within each pollutant section. Key performance characteristics are summarized through a common format. This information was developed from manufacturers' literature and conversations with their technical personnel. Additional information was gathered through reviewing research reports and journal articles cited in the summaries. For some instruments, entries for certain characteristics were left blank. These blank entries have occurred for the following reasons:

1. The performance characteristic does not apply. For example, zero drift for passive collection systems.

2. Potentially misleading assumptions are required. For example, lower limits of detection for collectors are influenced by choices in the analytical system. Given the sometimes broad range of available analytical devices and techniques, detection limits for such systems are not an intrinsic property of the collector alone.

3. Information was not available.

A relatively large number of stationary/active/analyzers are available for CO, NO_2, O_3, and SO_2. Commercially available units that appear in the EPA-designated list of reference and equivalent methods for these pollutants are listed separately in more condensed form at the end of this Appendix.

An explanation of instrument terms appears in Figure A-1. Pricing and manufacturers' information contained in the instrument summaries generally is as of 1985, and is subject to change. The instrument summaries were compiled for convenience of the reader; no product endorsement is implied.

| PRINCIPLE OF OPERATION | The chemical or physical basis for the measurement technique. |

| PERFORMANCE | Lower Detectable Limit: | The smallest quantity or concentration that causes a response equal to at least twice the noise level. |

Range: The lower and upper detectable limits; often synonymous with full scale.

Interferences: Substances or effects other than the measurement parameter that have been found to cause a measurable response in the instrument output. In many cases, tested interferents do not ordinarily occur indoors at concentrations high enough to affect instrument output.

Sampling Rate: Air volume flow through the instrument.

Accuracy: The percentage difference between true values that have been established by acceptable reference methods and measured values. Accuracy is generally referenced to full-scale reading of the output.

Reproducibility: The extent of variability among repeated measurements under the same input condition (pollutant concentration); often expressed as the standard deviation of the repeated measurements relative to the average measurement.

Zero Drift: The change in response to zero input or zero pollutant concentration over a specified period of continuous, unadjusted operation.

Span Drift: The change in response to an up-scale pollutant concentration over a specified period of continuous, unadjusted operation.

| OPERATION | Temperature Range: | The range of ambient temperatures over which the instrument meets or exceeds performance specifications. |

Relative Humidity Range: The range of ambient relative humidities over which the instrument meets or exceeds performance specifications.

Calibration: The manner in which instrument response is referenced to known standards.

Warmup Time: The amount of time required to achieve stable operation from initial startup.

(Continued)

FIGURE A-1. Explanation of instrument terms.

<u>Unattended Period</u>: The amount of time over which the instrument will meet or exceed performance specifications For many instruments summarized here, the principal limitation is battery life.

<u>Maintenance</u>: Service intervals recommended by the manufacturer.

<u>Power</u>: Specifications of alternating current and/or batteries required to operate the instrument.

FEATURES

<u>Output</u>: Defines the available readouts for the instrument. The cases for which laboratory analysis is necessary are denoted by "laboratory report."

<u>Training</u>: Defines the level of expertise required to operate (but not necessarily calibrate or repair) the instrument. Two categories of training are used: (1) "none required," implying that it can be successfully operated by nontechnical personnel with minimum instructions; (2) "recommended," implying that both a technical background and training are needed. For any given continuous analyzer or active collector, some degree of technical background is necessary for calibration, repair, and maintenance. For passive collectors, deployment can be carried out by nontechnical personnel with suitable instructions, but laboratory analysis always requires technical training.

<u>Options</u>: Additional features and accessories available from the manufacturer.

COSTS

Manufacturer-supplied costs in U.S. dollars.

MANUFACTURER

Mailing address, telephone number, and (where available) Telex or other special contact information.

REFERENCES

Published and other sources of information that were used to construct the summary. Separate listings are offered to denote specifications (generally manufacturer-supplied material) versus operations experience (user-supplied information).

REMARKS

Additional comments that are beyond the scope of the format used.

FIGURE A–1. Explanation of instrument terms. (Concluded)

ASBESTOS AND OTHER FIBROUS AEROSOLS
PORTABLE/ACTIVE/ANALYZER

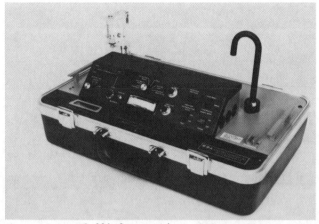

Photograph courtesy of GCA Corporation

Weight: 11.4 kg   Dimensions: 53 x 35 x 20 cm

PRINCIPLE OF
OPERATION

Induced oscillation/optical scattering. Sample air passes through a laminar flow chamber and enters a sensing region where an oscillating electric field induces fiber oscillations. The sensing region is illuminated by a continuous wave He-Ne laser that is aligned with sample flow. Scattering pulses from fiber oscillation are detected by a photomultiplier positioned at right angles to the laser. Electronic circuitry applies four separate acceptance tests to discriminate fibers, producing fiber counts per cubic centimeter.

PERFORMANCE

Lower Detectable Limit: 0.001 fibers/cm^3
                         minimum detectable fiber
                          length: 2 μm
                         minimum detectable fiber
                          diameter: 0.2 μm

Range: 0.001 to 30 fibers/cm^3

Interferences: large concentrations of elongated particles

Sampling Rate: 2 L/min (adjustable 1.5 to 2.5 L/min), continuous; fiber counting and selectable at 1, 10, 100, and 1 000 min

Accuracy: equal to reproducibility when calibrated for specific fibers

Reproducibility: (1σ) ± (100/N)%, where N is the number of fibers counted

Zero Drift:

Span Drift:

OPERATION

Temperature Range: 0 to 50 °C

Relative Humidity Range: 0 to 95% for conductive fibers
                          30 to 95% for dielectric fibers

Calibration: factory set; field adjustable through compar-
              ison with NIOSH asbestos fibers method (see
              Appendix B)

Warmup Time: 5 min

Unattended Period: indefinite

Maintenance: occasional cleaning of optics

Power: 115 or 220 V ac, 50 or 60 Hz; or may be run from
        battery power pack

FEATURES

Output: six-digit LCD, recorder output

Training: recommended

Options: battery power pack, digital to analog interface
          (recorder output)

COSTS

FAM-1: $10 850
Battery power pack: $720
Digital analog interface: $830

MANUFACTURER

GCA Corporation
Technology Division
213 Burlington Road
Bedford, Massachusetts   01730
(617) 275-5444

REFERENCES

A. Specifications:

    1. Manufacturer's bulletin--9-80 cp 2.5M.

    2. Lilienfeld, P., Development of a Prototype Fibrous
        Aerosol Monitor, Am. Ind. Hyg. Assoc. J., vol. 4,
        p. 270, 1979.

B. <u>Operations experience</u>:

 1. Elias, J.D., Dry Removal of Asbestos, <u>Am. Ind. Hyg.</u>
     <u>Assoc. J.</u>, 1981.

 2. Page, S.J., Correlation of the Fibrous Aerosol
     Monitor with the Optical Membrane Filter Count
     Technique, U.S. Department of the Interior, Bureau
     of Mines Report, 1980.

REMARKS   1. A standard inline membrane filter permits concurrent
                         collection of fiber samples.

NOTES

BIOLOGICAL AEROSOLS
STATIONARY/ACTIVE/COLLECTOR

B-1
Andersen
#10-800 Viable
 Sample Kit
1 of 2

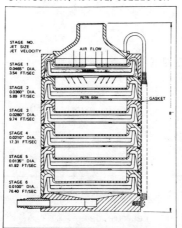

Illustration courtesy of Andersen Samplers, Inc.

Weight: 2.5 kg   Dimensions: 20 x 11 cm

PRINCIPLE OF OPERATION	Impaction. Upon entering the inlet, sample air is accelerated through a series of six impaction stages, each of which holds a petri dish containing agar, which serves as the collection surface. Within each stage, jet velocity is uniform but increases in each succeeding stage. Each successive stage collects smaller particles. Microbial colonies are incubated for 24 h and counted manually.

PERFORMANCE

Lower Detectable Limit:

Range:

Interferences:

Sampling Rate: 28.3 L/min, continuous

Accuracy:

Reproducibility:

Zero Drift:

Span Drift:

OPERATION

Temperature Range:

Relative Humidity Range:

B-1
Andersen
#10-800 Viable
 Sample Kit
2 of 2

Calibration: see remark 2

Warmup Time:

Unattended Period: usually <60 min (see remark 1)

Maintenance:

Power: 115 V ac

FEATURES   Output: laboratory report

Training: recommended

Options:

COSTS $2 195

MANUFACTURER   Andersen Samplers, Inc.
4215 Wendell Drive
Atlanta, Georgia 30336
(404) 691-1910
(800) 241-6898

REFERENCES   A. Specifications:

1. Manufacturer's bulletin.

B. Operations experience:

1. Chatigny, M.A., Sampling Airborne Microorganisms, in
Air Sampling Instruments for Evaluation of Atmospheric Contaminants, eds. P.J. Lioy and M.J.Y. Lioy
(6th edition), American Conference of Governmental
Industrial Hygienists, Cincinnati, 1983.

2. Solomon, W.R., Sampling Technique for Airborne
Fungi, in Mould Allergy, eds. Y. Al-Doory and
J.F. Domson, Lea and Febiger, Philadelphia, 1984.

REMARKS   1. Sample periods for biological aerosols are generally
short--a few minutes at most--to avoid dehydration of
collected microorganisms.

2. Flow rate is controlled by an adjustable valve on the
pump; periodic calibration is recommended.

BIOLOGICAL AEROSOLS
STATIONARY/ACTIVE/COLLECTOR

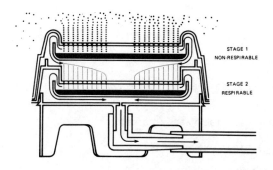

STAGE 1
NON-RESPIRABLE

STAGE 2
RESPIRABLE

Illustration courtesy of Andersen Samplers, Inc.

Weight: 1.5 kg   Dimensions: 10 x 12 cm

PRINCIPLE OF   Impaction. Upon entering the inlet, sample air is acceler-
OPERATION   ated through a series of two impaction stages; each stage
                   holds a disposable petri dish containing agar, which serves
                   as the collection surface. The first stage collects parti-
                   cles larger than 7 µm; the second stage collects particles
                   between 1 and 7 µm. Microbial colonies are incubated for
                   24 h and counted manually.

PERFORMANCE   Lower Detectable Limit:

                   Range:

                   Interferences:

                   Sampling Rate: 28.3 L/min, continuous

                   Accuracy:

                   Reproducibility:

                   Zero Drift:

                   Span Drift:

OPERATION   Temperature Range:

                   Relative Humidity Range:

                   Calibration: none required in ordinary use (see remark 1)

                   Warmup Time:

Unattended Period: usually <60 min (see remark 2)

Maintenance:

Power: defined by user-supplied vacuum source

FEATURES

Output: laboratory report

Training: recommended

Options:

COSTS

$850

MANUFACTURER

Andersen Samplers, Inc.
4215 Wendell Drive
Atlanta, Georgia 30336
(404) 691-1910
(800) 241-6898

REFERENCES

A. Specifications:

  1. Manufacturer's bulletin.

B. Operations experience:

  1. Chatigny, M.A., Sampling Airborne Microorganisms, in Air Sampling Instruments for Evaluation of Atmospheric Contaminants, eds. P.J. Lioy and M.J.Y. Lioy (6th edition), American Conference of Governmental Industrial Hygienists, Cincinnati, 1983.

  2. Solomon, W.R., Sampling Technique for Airborne Fungi, in Mould Allergy, eds. Y. Al-Doory and J.F. Domson, Lea and Febiger, Philadelphia, 1984.

REMARKS

1. A critical orifice situated in the base of the sampler provides constant flow of 1 ft^3/m as long as vacuum is 10 in of Hg.

2. Sample periods for biological aerosols are generally short--a few minutes at most--to preclude dehydration of collected microorganisms.

3. The sampler uses disposable 100-mm petri dishes and is reusable and sterilizable.

CARBON MONOXIDE
PORTABLE/ACTIVE/ANALYZER

Photograph courtesy of Energetics Science, Inc.

Weight: 4.5 kg   Dimensions: 17.8 x 17.8 x 33 cm

PRINCIPLE OF OPERATION	Electrochemical oxidation. Ambient air is drawn past a catalytically active electrode where CO is oxidized, producing a signal proportional to the CO concentration in the sample airstream. Potential interferents can be removed by an inlet scrubber.

PERFORMANCE

Lower Detectable Limit: <0.5 ppm

Range: 0 to 100 ppm, 0 to 600 ppm

Interferences: expressed as ppm of interferent needed to give 1 ppm deflection (testing performed with Purafil filter installed): CH_2 (acetylene) = 0.3, C_2H_4 (ethylene) = 1.0, C_2H_6 (ethane) = 100, C_3H_8 (propane) = 100, H_2 = 100, H_2S = 100, SO_2 = 100

Sampling Rate: 700 mL/min, continuous

Accuracy: ±1%

Reproducibility: ±1%

Zero Drift: ±0.5 ppm/d

Span Drift: 1%/d

OPERATION	Temperature Range: 0 to 40 °C
	Relative Humidity Range: unaffected by water vapor
	Calibration: standard gas mixture
	Warmup Time: specified only as "brief"
	Unattended Period: 8+ h on battery
	Maintenance: 1-yr sensor warranty; boards replaceable in field
	Power: 105-125 V ac at 50-60 Hz; Ni-Cd batteries with built-in recharger
FEATURES	Output: panel meter with parallax mirror; 0- to 1-V dc recorder output
	Training: none required for sampling
	Options: dc-powered recorder, ac-powered recorder
COSTS	2000 Series CO monitor: $1 900
	dc recorder: $550
	ac recorder: $450
MANUFACTURER	Energetics Science, Inc.
	6 Skyline Drive
	Hawthorne, New York 10532
	(914) 592-3010
REFERENCES	A. Specifications:
	1. Manufacturer's bulletin.
	B. Operations experience:
	1. Cortese, A.D., and Spengler, J.D., Ability of Fixed Monitoring Station to Represent Personal Carbon Monoxide Exposures, J. Air Pollut. Control Assoc., vol. 26, pp. 1144-1150, 1976.

C-1
Energetics Science
2000 Series
CO Analyzer
3 of 3

REMARKS

1. Low temperature (0 to 10 °C) zero drift was found to be +1% to 2% of scale over 30 min; calibration drift was <1 ppm (Cortese and Spengler 1976).

2. The model 7000 version of the instrument features two transducers (selected from CO, NO_2, NO, and H_2S). Although it would be attractive to pair CO and NO_2 for indoor air quality monitoring, this may be of limited use because the most sensitive range for NO_2 is 0 to 2 ppm.

NOTES

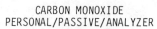

CARBON MONOXIDE
PERSONAL/PASSIVE/ANALYZER

C-2
Energetics Science
Model 210
Personal CO Monitor
1 of 2

Photograph courtesy of Energetics Science, Inc.

Weight: 0.3 kg   Dimensions: 14 x 8.5 x 3.8 cm

PRINCIPLE OF OPERATION	Electrochemistry. Ambient air diffuses into a patented, three-electrode electrochemical cell.

PERFORMANCE

Lower Detectable Limit: 1 ppm

Range: 0 to 1 999 ppm

Interferences: expressed as ppm of interferent needed to give 1 ppm deflection: NH_3 = 135, NO_2 = 270, SO_2 = 145, H_2S = 130, C_2H_2 (acetylene) = 170, C_2H_4 (ethylene) = 135, C_2H_6 (ethane) = 1 200, C_2H_8 (propane) = 425, C_2H_5OH (ethanol) = 140, 2-propanol = 750

Sampling Rate: diffusion, continuous

Accuracy: ±5% or ±1 ppm (whichever is greater)

Reproducibility: ±2% or ±1 ppm (whichever is greater)

Zero Drift: <5 ppm/24 h

Span Drift: ±2%/24 h or 2 ppm/24 h (whichever is greater)

OPERATION

Temperature Range: 0 to 40 °C

Relative Humidity Range: 5 to 90%

C-2
Energetics Science
Model 210
Personal CO Monitor
2 of 2

Calibration: standard gas mixture

Warmup Time:

Unattended Period:

Maintenance: batteries are field replaceable, 6-mo sensor
warranty

Power: standard 9-V transistor battery

FEATURES	Output: LCD push-button-activated or continuous display
	Training: none required for sampling
	Options: lapel clip-on alarm horn for high noise areas

COSTS	$695

MANUFACTURER	Energetics Science, Inc. 6 Skyline Drive Hawthorne, New York 10532 (914) 592-3010

REFERENCES	A. Specifications:
	1. Manufacturer's brochure 2C-10-1-82.
	B. Operations experience:

REMARKS	1. A similar version is available for H_2S.

CARBON MONOXIDE
PERSONAL/ACTIVE/ANALYZER

C-3
General Electric
CO Detector
1 of 3

Photograph courtesy of General Electric Company

Weight: 0.3 kg   Dimensions: 7.5 x 13.5 x 3.6 cm

PRINCIPLE OF
OPERATION

Electrochemistry. Air is drawn through a filter and into an electrochemical cell in which oxidation of CO produces an electrical signal proportional to CO concentration in the airstream.

PERFORMANCE

Lower Detectable Limit: 1 ppm

Range: 0 to 1 000 ppm

Interferences: expressed as ppm of interferent needed to give 1 ppm deflection (testing performed with Purafil filter installed): H_2 = 50, C_2H_2 (acetylene) = 6, C_2H_4 (ethylene) = 6; electromagnetic fields (3 V/m at 0 to 80 MHz) showed no effect

Sampling Rate: 60 mL/min, continuous

Accuracy: direct LCD readout

Reproducibility: ±5%

Repeatability: ±5%

Zero Drift: very little (usually ±1 ppm over several days)

Span Drift: generally ±5 ppm at 60 ppm span gas if several days elapse

OPERATION	Temperature Range: 1 to 40 °C (freezing conditions should be avoided)
	Relative Humidity Range: 0 to 95%
	Calibration: standard gas mixture
	Warmup Time: 3 min
	Unattended Period: 10 h on battery, unlimited on ac power
	Maintenance: Purafil filter: renew upon color change cell assembly: replenish distilled or deionized water periodically storage conditions: 1 to 50 °C
	Power: 5.2 V dc, 250 mA-h, rechargeable Ni-Cd batteries
FEATURES	Output: LCD panel readout (instantaneous levels), recorder output 0 to 1 V dc, internal accumulator (requires external console to read out; see options)
	Training: none required for sampling
	Options: support console (to read/reset accumulator), gas calibration kit charger
COSTS	Direct indicating detector: $1 195 Support console: $775 (single charge), $935 (multicharge) Gas calibration kit: $245 Charger: $29
MANUFACTURER	General Electric Company 333 West Seymour Avenue Cincinnati, Ohio 45216 (513) 948-5050
REFERENCES	A. Specifications: 1. Operation and Maintenance Instructions, Direct Indicating SPE Carbon Monoxide Detector. GE Aircraft Equipment Devices, 1980.

2. Model 15ECS1CO2 Carbon Monoxide Dosimeter and Model 15ECS3CO3 Direct Indicating Carbon Monoxide Detector for Performance and Intrinsically Safe for Classes I and II, Divisions 1 and 2, Groups A, B, C, D, E, F, and G Hazardous Locations, J.I. 1A7AO, Ax (6340/3610), Factory Mutual.

B. Operations experience:

1. Akland, G.G., Hartwell, T.D., Johnson, T.R., and Whitmore, R.W., Measuring Human Exposure to Carbon Monoxide in Washington, D.C., and Denver, Colorado, during the Winter of 1982-1983, Environ. Sci. Technol., vol. 19, no. 10, pp. 911-918, 1985.

2. Nagda, N.L., and Koontz, M.D., Microenvironmental and Total Exposures to Carbon Monoxide for Three Population Subgroups, J. Air Pollut. Control Assoc., vol. 35, no. 2, pp. 134-137, 1985.

REMARKS

1. These units may be leased from the manufacturer.

2. The unattended period has been extended to well over 35 h by substituting a larger capacity battery (Nagda and Koontz 1985).

3. These units have been approved by the following organizations:

   Mine Safety and Health Administration, U.S. Department of Labor. Permissible Carbon Monoxide Detector, Tested in Methane--Air Mixtures Only, Approval 2G-3152-1.

   Factory Mutual System. Approved for Performance and Intrinsically Safe for Classes I and II, Divisions 1 and 2, Groups A, B, C, D, E, F, and G.

NOTES

CARBON MONOXIDE
POR TABLE/ACTIVE/ANALYZER

C-4
Interscan
Models 1140 and
 4140
CO Analyzers
1 of 3

Photograph courtesy of Interscan Corporation

Weight: 3.6 kg (1140)   Dimensions: 18 x 15 x 29 cm (Model 1140)
          2.0 kg (4140)   18 x 10 x 23 cm (Model 4140)

PRINCIPLE OF   Electrochemistry. Gas molecules from the moving sample
OPERATION   airstream pass through a diffusion medium and are adsorbed
                    onto an electrocatalytic sensing electrode where subsequent
                    reactions generate an electric current. The diffusion-
                    limited current is linearly proportional to the CO
                    concentration.

PERFORMANCE   Lower Detectable Limit: 1% of full scale

                    Range: 0 to 100 ppm, 0 to 250 ppm, 0 to 500 ppm (other
                             ranges available)

                    Interferences: expressed as ppm of interferent needed to
                                     give 1 ppm deflection: $H_2 = 100$, $NH_3 = 5$.
                                     A gas scrubber is required when unsaturated
                                     hydrocarbons are present in concentrations
                                     equivalent to CO.

                    Sampling Rate: 1.2 L/min, continuous

                    Accuracy: ±2% of full scale

                    Reproducibility: ±0.5% of full scale

                    Zero Drift: ±1% full scale in 24 h

                    Span Drift: <±2% full scale in 24 h

C-4
Interscan
Models 1140 and
 4140
CO Analyzers
2 of 3

OPERATION	Temperature Range: 10 to 120 °F
	Relative Humidity Range: 1 to 100%
	Calibration: standard gas mixture
	Warmup Time: <5 min
	Unattended Period: 10 h on battery power
	Maintenance: calibration, battery replacement, biannual sensor replacement
	Power: 1140: four alkaline MnO_2 batteries for amplifier, two Ni-Cd for pumps and power-on LED, one HgO battery for bias amplifier reference
	4140: four "1/2C" Ni-Cd batteries
FEATURES	Output: 0-100 mV full scale
	Training: none required for sampling
	Options: alarms, special ranges
COSTS	1140: $1 675 4140: $1 895
MANUFACTURER	Interscan Corporation P.O. Box 2496 21700 Nordhoff Street Chatsworth, California 91311 (818) 882-2331 TELEX: 67-4897
REFERENCES	A. Specifications: 1. Manufacturer's bulletin.

C-4
Interscan
Models 1140 and
 4140
CO Analyzers
3 of 3

B. Operations experience:

1. Ziskind, R.A., Rogozen, M.B., Carlin, T., Drago, R., Carbon Monoxide Intrusion into Sustained-Use Vehicles, Environ. Int., vol. 5, pp. 109-123, 1981.

REMARKS

NOTES

CARBON MONOXIDE
PERSONAL/PASSIVE/ANALYZER

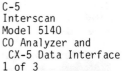

C-5
Interscan
Model 5140
CO Analyzer and
 CX-5 Data Interface
1 of 3

Photograph courtesy of Interscan Corporation

Weight: 0.68 kg (5140)
         0.61 kg (CX-5)

Dimensions: 15 x 8 x 5 cm (5140)
             7 x 16 x 3 cm (CX-5)

PRINCIPLE OF OPERATION	Diffusion/electrochemistry. CO diffuses into an electro-chemical cell, producing a signal proportional to the CO concentration. The signal is digitized, incorporated into 1-min averages, and stored. Nondestructive recovery of each 1-min average is accomplished through a separate data reader. Data storage capacity is 2 048 1-min averages. Stored data can be retrieved through the CX-5 interface and transferred to a computer for further processing.
PERFORMANCE	Lower Detectable Limit: 0.5% of full scale (2.5 ppm)

Range: 0 to 500 ppm

Interferences: expressed as ppm of interferent needed to give 1 ppm deflection: H_2 = 100, NH_3 = 5

Sampling Rate: diffusion, continuous

Accuracy: ±2% of reading, ±1 Least Significant Digit (LSD), ±0.5% of full scale

Reproducibility: ±1% reading, ±1 LSD

Zero Drift: ±1% reading, ±1 LSD in 24 h

Span Drift: ±1% reading, ±1 LSD in 24 h |

C-5
Interscan
Model 5140
CO Analyzer and
 CX-5 Data Interface
2 of 3

OPERATION	Temperature Range: 30 to 120 °F
	Relative Humidity Range: 1 to 100%
	Calibration: standard gas mixture
	Warmup Time: <5 min
	Unattended Period: up to 34 h
	Maintenance: calibration, battery replacement, sensor replacement
	Power: long-life 9-V battery (alkaline MnO_2 NEDA type 1604A); battery life is 125 h (continuous operation)
FEATURES	Output: 8-bit parallel or serial (RS232C) (see remarks 1 and 2)
	Training: none required for sampling
	Options:
COSTS	5140: $1 145
CX-5: $1 375
MANUFACTURER	Interscan Corporation
P.O. Box 2496
21700 Nordhoff Street
Chatsworth, California 91311
(213) 882-2331
TELEX: 67-4897
REFERENCES:	A. Specifications:
	1. Manufacturer's bulletin.
	B. Operations experience:
	1. M. Shaw, Time-History Toxic Gas Dosimetry, Ind. Hyg. News, September, pp. 45-47, 1983.

C-5
Interscan
Model 5140
CO Analyzer and
 CX-5 Data Interface
3 of 3

REMARKS

1. The CX-5 interface allows nondestructive retrieval of all 1-min averages stored in the dosimeter memory. The interface is programmed to give time-selected and time-weighted averages, maximum short-term exposure levels, and the time at which they occur, as well as the peak concentration and the time at which it occurs. One CX-5 can service many 5000 series dosimeters.

2. Data readout may also be accomplished by a device available from:

   Metrosonics, Inc.
   P.O. Box 23075
   Rochester, New York 14692
   (716) 334-7300

3. Interscan also produces the Model 2140 CO personal monitor that offers an LCD display of concentration instead of data logging.

NOTES

FORMALDEHYDE
PERSONAL/PASSIVE/COLLECTOR

Photograph courtesy of Air Quality Research, Inc., International

Weight: negligible   Dimensions: 9-cm length, 2.5-cm diameter

---

PRINCIPLE OF
OPERATION

Molecular diffusion/sorption. The sampler consists of a
glass-fiber filter treated with sodium bisulfite, housed in
a glass vial that is capped when not in use. HCHO diffuses
through the tube at a rate dependent on Fick's First Law of
Diffusion. The treated filter at the bottom end of the tube
maintains a near-zero HCHO concentration at the base; therefore, the quantity of HCHO transferred through the diffusion
path is related to the ambient concentration and the length
of time exposed. Collected HCHO is quantified in the
laboratory using the chromotropic acid procedure.

PERFORMANCE

Lower Detectable Limit: 1.68 ppm h (0.010 ppm for 1-week
                           exposure)

Range: validated over the range of 0 to 150 ppm h; capacity
        is in excess of 1 000 ppm h

Interferences: none known at this time. The analytical
                procedure (chromotropic acid) is subject to
                interference by several compounds, but they
                are seldom encountered in indoor air quality
                sampling applications. In any event, the
                compounds are not expected to be collected
                by the bisulfite-treated filter collection
                element.

F-1
Air Quality Research,
 Inc., International
PF-1
HCHO Passive Monitor
2 of 3

Sampling Rate: diffusion, continuous

Accuracy:

Reproducibility: ±15%

Zero Drift:

Span Drift:

OPERATION

Temperature Range: 15 to 35 °C

Relative Humidity Range: noncondensing

Calibration: laboratory standards

Warmup Time:

Unattended Period: 1 week (recommended minimum exposure for
indoor studies)

Maintenance: shelf life has been validated for 6 mo

Power: none required for sampling

FEATURES

Output: laboratory report

Training: none required for sampling

Options:

COSTS

Sampler only: $15 for box of two
Sampler plus analysis by AQRI: $30 for box of two

NOTE: These are nominal prices; actual costs depend upon
lot sizes.

MANUFACTURER

Air Quality Research, Inc., International
901 Grayson Street
Berkeley, California 94710
(415) 644-2097

REFERENCES

A. Specifications:

1. Manufacturer's bulletin.

2. Geisling, K.L., Tashima, M.K., Girman, J.R.,
    Miksh, R.R., A New Passive Monitor for Determin-
    ing Formaldehyde in Indoor Air, <u>Environ. Int.</u>,
    vol. 8, pp. 153-158, 1982.

3. National Institute for Occupational Safety and
    Health, <u>Man. Anal. Methods</u>, 2d ed., vol. 1,
    pp. 125-1 to 125-9.

B. <u>Operations experience</u>:

1. Sexton, K., Liu, K., Petreas, M.X., "Measuring
    Indoor Air Quality by Mail," <u>Proceedings of the
    78th Annual Meeting of the Air Pollution Control
    Association</u>, Paper no. 85-31.3, Pittsburgh,
    1985.

REMARKS   1. These devices do not require specialized training for
                       use. However, extreme care must be exercised in proper
                       placement in the field and in recordation of the
                       exposure interval. The units should be exposed at least
                       in immediate pairs at each sampling point. Therefore, a
                       simple indoor/outdoor comparison, for instance, would
                       require four samplers.

NOTES

FORMALDEHYDE
STATIONARY/ACTIVE/ANALYZER

F-2
CEA
TGM 555
Formaldehyde Analyzer
1 of 2

Photograph courtesy of CEA Instruments, Inc.

Weight: 14 kg   Dimensions: 51 x 41 x 18 cm

PRINCIPLE OF OPERATION	Automated wet chemistry/colorimetry. Sample air is drawn through a sodium tetrachloromercurate solution that contains a fixed quantity of sodium sulfite. Acid bleached pararosaniline is added, and the intensity of the resultant color is measured at 550 nm. Reagent handling and processing is automatic.

PERFORMANCE

Lower Detectable Limit: 0.002 ppm

Range: 0 to 5 ppm (adjustable from 0 to 0.15 up to
         0 to 10 ppm)

Interferences: none

Sampling Rate: 500 mL/min, continuous

Accuracy: ±3% (referenced to chromotropic acid procedure)

Reproducibility: 1%

Zero Drift: <2% in 24 h

Span Drift: <2% in 24 h

OPERATION

Temperature Range: 16 to 27 °C optimum; 4 to 50 °C usable

Relative Humidity Range: 5 to 95%

    Calibration: with liquid standards or HCHO permeation tubes

    Warmup Time: 20 min

    Unattended Period: 18 h on fully charged batteries

    Maintenance: tubing in the peristaltic pump should be
              changed once a month

    Power: 12 V dc unregulated, 4 W; 115/230 V ac, 50/60 Hz

FEATURES

    Output: digital panel meter
           0 to 1 V at 0 to 2.0 mA recorder output

    Training: none required for sampling

    Options: stream splitter (to multiply range)

COSTS

    TGM 555: $5 410
    Stream Splitter: $295

MANUFACTURER

    CEA Instruments, Inc.
    16 Chestnut Street
    P.O. Box 303
    Emerson, New Jersey 07630
    (201) 967-5660

REFERENCES

    A. Specifications:

        1. Manufacturer's bulletin.

    B. Operations experience:

        1. Matthews, T.E., and Howell, T.C., Visual Colori-
           metric Formaldehyde Screening Analysis for Indoor
           Air, J. Air Pollut. Control Assoc., vol. 31,
           pp. 1181-1184, 1981.

        2. Balmat, J.L., and Meadows, G.W., Monitoring
           Formaldehyde in Air, Am. Ind. Hyg. Assoc. J.,
           vol. 46, no. 10, pp. 578-584, 1985.

REMARKS

    1. A calibration gas generator, the SC-100, which operates
        at 100 °C using HCHO permeation tubes, is available for
        dynamic calibration of the TGM 555.

F-3
Du Pont
PRO-TEK™
HCHO Passive
 Dosimeter
Type C60
1 of 3

FORMALDEHYDE
PERSONAL/PASSIVE/COLLECTOR

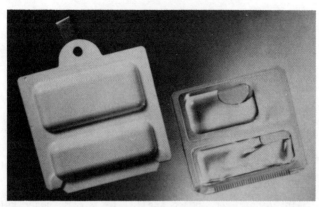

Photograph courtesy of Du Pont Applied Technology Instruments

Weight: negligible   Dimensions: 7.6 x 7.1 x 0.9 cm

PRINCIPLE OF OPERATION	Molecular diffusion/sorption. Collection relies on molecular diffusion to deliver sample air to a liquid sorbent solution at a constant rate. After exposure, the sorbent is analyzed in a laboratory spectrophotometer for HCHO content and the time-weighted average concentration.

PERFORMANCE

Lower Detectable Limit: 1.6 ppm h (0.010 ppm for 1-week exposure)

Range: 1.6 to 54 ppm h

Interferences: none

Sampling Rate: diffusion, continuous

Accuracy: ±13.1% (overall system accuracy) over the range of 1.6 to 54 ppm h

Reproducibility: ±5.9%

Zero Drift:

Span Drift:

F-3
Du Pont
PRO-TEK™
HCHO Passive
 Dosimeter
Type C60
2 of 3

OPERATION	Temperature Range: 4 to 49 °C
	Relative Humidity Range:
	Calibration: laboratory standards
	Warmup Time:
	Unattended Period: 2 to 18 h
	Maintenance: unexposed shelf life is 6 mo under refriger-ation, 3 mo at room temperature; shelf life of exposed badges is 2 weeks
	Power: none required for sampling
FEATURES	Output: laboratory report
	Training: see remark 2
	Options:
COSTS	Type C60, 10 per box:
	1-10 boxes: $222 (Order Code 5147)
	11-25 boxes: $201 (Order Code 5148)
	26+ boxes: $160 (Order Code 5149)
MANUFACTURER	Du Pont Applied Technology Instruments
	Box 10
	Kennett Square, Pennsylvania 19348
	(215) 444-4493
	(800) 344-4900
REFERENCES	A. Specifications:
	1. Manufacturer's sampling and analytical procedure.
	2. Kring, E.V., Thornley, G.D., Dessenberger, C., Lautenberger, W.J., and Ansul, G.R., A New Passive Colorimetric Air Monitoring Badge for Sampling Formaldehyde in Air, Am. Ind. Hyg. Assoc. J., vol. 43, pp. 786-795, 1982.

F-3
Du Pont
PRO-TEK™
HCHO Passive
 Dosimeter
Type C60
3 of 3

B. Operations experience:

1. Kring, E.V., Ansul, G.R., Basilio, A.M., Jr.,
    McGibney, P.D., Stephens, J.S., and O'Dell, H.L.,
    Sampling for Formaldehyde in Workplace and Ambient
    Air Environments--Additional Laboratory Validation
    and Field Verification of a Passive Monitoring
    Device Compared with Conventional Sampling Methods,
    Am. Ind. Hyg. Assoc. J., vol. 45, pp. 318-324, 1984.

2. Balmat, J.L., and Meadows, G.W., Monitoring
    Formaldehyde in Air, Am. Ind. Hyg. J., vol. 46,
    no. 10, pp. 578-584, 1985.

3. Coyne, L.B., Cook, R.E., Mann, J.R., Bouyoucos, S.,
    McDonald, O.F., and Baldwin, C.L., Formaldehyde:
    A Comparative Evaluation of Four Monitoring Methods,
    Am. Ind. Hyg. Assoc. J., vol. 46, no. 10,
    pp. 609-619, 1985.

REMARKS

1. A maximum exposure time has not been defined; laboratory
    validation exposure times varied from 2 to 18 h. The
    shelf life of exposed badges is 2 weeks, suggesting this
    to be a maximum exposure time.

2. Each badge carries two compartments of sorbent solution;
    one is used for sampling, the other (thoroughly sealed
    until analysis) acts as a blank.

3. Du Pont does not plan to market this device directly to
    homeowners because of the absence of professional super-
    vision to ensure accuracy for sampling results.

4. An analytical service for exposed badges is available
    from a number of AIHA-accredited laboratories.

NOTES

F-4
3M
Formaldehyde
 Monitor 3750
1 of 2

FORMALDEHYDE
PERSONAL/PASSIVE/COLLECTOR

Photograph courtesy of 3M Company

      Weight: negligible   Dimensions:

PRINCIPLE OF OPERATION

Molecular diffusion/sorption. HCHO diffuses into the monitor and is collected by a chemisorption process onto an impregnated media. At a constant sampling rate, the amount of HCHO adsorbed is controlled by concentration and exposure time. At the end of sampling, the monitor is sealed and taken to the laboratory where collected HCHO is desorbed using water and quantitated spectrophotometrically. The weight of recovered HCHO is linearly related to the time-weighted-average exposure.

PERFORMANCE

Lower Detectable Limit: 0.8 ppm h (0.005 ppm for 1-week exposure)

Range: up to 72 ppm h

Interferences: phenol, alcohols, and unsaturated compounds at 10 to 20 times the HCHO concentration

Sampling Rate: 4.88 µg/ppm h, continuous

Accuracy: <±25%; exceeds OSHA accuracy requirements

Reproducibility:

Zero Drift:

Span Drift:

F-4
3M
Formaldehyde
 Monitor 3750
2 of 2

OPERATION	Temperature Range: -20 to 130 °F
	Relative Humidity Range: 15 to 95%
	Calibration: laboratory standards
	Warmup Time:
	Unattended Period: up to 1 week
	Maintenance: shelf life of unexposed samplers is 1 yr at room temperature; exposed samples have a shelf life up to 4 weeks at room temperature
	Power: none required for sampling
FEATURES	Output: laboratory report
	Training: none required for sampling
	Options:
COSTS	3750 (sampler plus analysis at 3M): $35 3751 (sampler only): $21
MANUFACTURER	Occupational Health and Safety Products Division/3M 220-7W, 3M Center St. Paul, Minnesota 55144 (612) 733-6234
REFERENCES	A. Specifications:
	1. Manufacturer's brochure.
	2. Rodriguez, S.T., Olsen, P.B., and Lund, V.R., Colorimetric Analysis of Formaldehyde Collected on a Diffusional Monitor, Technical Bulletin R-AIHA5(71.1)R, 3M Company, St. Paul, 1981.
	B. Operations experience:
REMARKS	

INHALABLE PARTICULATE MATTER
PERSONAL/PASSIVE/ANALYZER

I-1
GCA Corporation
MINIRAM
Aerosol Monitor
1 of 3

Photograph courtesy of GCA Corporation

Weight: 0.4 kg   Dimensions: 10 x 10 x 4 cm

PRINCIPLE OF OPERATION	Optical scattering. Sample air passes through the open sensing volume by free convection. A silicon detector with an interference filter senses forward light scattering from a pulsed near-infrared-emitting diode.

PERFORMANCE

Lower Detectable Limit: scattering coefficient of approximately 10^{-5} m^{-1} (mass equivalence is tied to reference dust)

Range: 0.01 to 100 mg/m^3 (autoranging 0-10, 0-100)

Interferences: extreme fluctuations in ambient light

Sampling Rate: open convection, continuous

Accuracy: if calibrated for specific aerosol, equal to reproducibility

Reproducibility:

±0.05 mg/m^3 for 10-s measurement
±0.02 mg/m^3 for 1-min average
±0.006 mg/m^3 for 10-min average
±0.003 mg/m^3 for 1-h average

Zero Drift:

Span Drift:

I-1
GCA Corporation
MINIRAM
Aerosol Monitor
2 of 3

OPERATION	Temperature Range: 0 to 50 °C
	Relative Humidity Range: 0 to 95%
	Calibration: automatic zero reference in clean environment, optional reference scatterer, or gravimetric reference calibration
	Warmup Time: 1 min
	Unattended Period: at least 8.5 h with battery
	Maintenance: occasional cleaning or replacement of slide-in sensing chamber
	Power: internal rechargeable 7.5-V battery; charger operates from ac line
FEATURES	Output: three-digit LCD (updated every 10 s) 0- to 2-V analog recorder output; digital
	Training: none required for sampling
	Options: miniature strip chart recorder, zero check filter air unit, personal filter sample adaptor, respirator/face mask monitoring adaptor, shoulder strap, table stand, extra battery pack
COSTS	MINIRAM: $1 445 (includes charger/ac line adaptor, instrument/accessory case, manual)
	Recorder: $1 170
	Personal sampler adaptor: $250
	Respirator adaptor: $150
	Zero check unit: $260
	Shoulder strap: $25
	Table stand: $25
MANUFACTURER	GCA Corporation Technology Division 213 Burlington Road Bedford, Massachusetts 01730 (617) 275-5444

REFERENCES

A. Specifications:

1. Manufacturer's bulletin.

2. Lilienfeld, P., Final Report to the Bureau of Mines on Contract No. H0308132, 1982.

3. Lilienfeld, P., Current Mine Dust Monitoring Instrumentation Developments, Proceedings of the 1981 International Symposium on Aerosols in the Mining and Industrial Work Environment, 1983.

B. Operations experience:

1. Nagda, N.L., Fortmann, R.C., Koontz, M.D., and Rector, H.E., Comparison of Instrumentation for Microenvironmental Monitoring of Respirable Particulates, Proceedings of the 78th Annual Meeting of the Air Pollution Control Association, Paper No. 85-30A.1, Pittsburgh, 1985.

REMARKS

1. The unit comes with a factory calibration based on a representative test dust. An internal control allows adjustment of response to match any reference gravimetric calibration.

2. The GCA MINIRAM offers the following data handling capabilities:

   Readouts: selectable, 10-s measurements, time-averaged measurements, shift-averaged measurements, elapsed sampling time

   Storage: seven average concentrations, sampling times, off-times, and sampler identification number

   Memory playback: either through instrument's own LCD or by 300-baud ASCII (20 mA loop or RS232 may be connected with proper interface).

NOTES

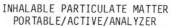

INHALABLE PARTICULATE MATTER
PORTABLE/ACTIVE/ANALYZER

I-2
GCA Corporation
RAM-1
Aerosol Monitor
1 of 3

Photograph courtesy of GCA Corporation

Weight: 4 kg   Dimensions: 20 x 20 x 20 cm

PRINCIPLE OF
OPERATION

Optical scattering. As sample air, drawn by pump, passes through the sensing volume, a silicon detector with an interference filter senses forward light scattering from a pulsed near-infrared-emitting diode. The upper limit of the particle size range is 20 µm; a series of precollectors offer cutpoints of 1, 2, 4, and 8 µm.

PERFORMANCE

Lower Detectable Limit: Scattering coefficient of approximately 4×10^{-6} m^{-1} (mass equivalence is tied to reference dust)

Range: 0.001 to 200 mg/m^3 (selectable 0 to 2, 0 to 20, or 0 to 200)

Interferences: none

Sampling Rate: 2 L/min (adjustable 1 to 3 L/min), continuous

Accuracy: if calibrated for specific aerosol, equal to reproducibility

Reproducibility: ±0.1% of full scale or ±0.005 mg/m^3 (whichever is larger)

Zero Drift: ±0.1% or ±0.005 mg/m^3

Span Drift: determined by measurement and zero precision stability over 24 h (whichever is larger)

I-2
GCA Corporation
RAM-1
Aerosol Monitor
2 of 3

OPERATION	Temperature Range: 0 to 50 °C
	Relative Humidity Range: 0 to 95%
	Calibration: reference scatterer or gravimetric reference calibration
	Warmup Time: <1 sec
	Unattended Period: at least 6 h on battery, unlimited on charger
	Maintenance: refillable diffusion-type drying cartridge for use in condensing atmospheres; high capacity filter cartridges externally accessible
	Power: internal rechargeable 6-V battery; charger operates from ac line
FEATURES	Output: four-digit LCD (updated three times each s); 0- to 10-V ac recorder output (minimum load impedance: 1 000 ohms)
	Training: none required for sampling
	Options: miniature strip chart recorder; intrinsic safety version available; averager/integrator
COSTS	RAM-1: $5 950 (includes charger/ac line adaptor, charger cable, cyclone preselector, inlet flow restrictor, two replacement filter cartridges, refillable desiccator, carrying strap, instrument/accessory case, manual)
	Recorder: $1 170
	Intrinsic safety version: $6 550
	Averager/integrator: $1 490
MANUFACTURER	GCA Corporation Technology Division
	213 Burlington Road
	Bedford, Massachusetts 01730
	(617) 275-5444
REFERENCES	A. Specifications:
	1. Manufacturer's bulletin, #2-80 CP/5M.

2. Tomb, T.F., Treaftis, H.N., and Gero, A.J.,
Instantaneous Dust Exposure Monitors, _Environ._
Int., vol. 5, pp. 85-96, 1981.

B. Operations experience:

1. Chansky, S.H., Lilienfeld, P., and Wiltsee, K.,
Evaluation of GCA Corporation's Model RAM-S as
an Equivalent Alternative to the Vertical Elutriator
for Cotton Dust Measurement. Natural Fibers Textile
Conference, Charlotte, 1979.

2. Rubow, K.L., and Marple, V.A., An Instrument
Evaluation Chamber, Calibration of Commercial
Photometers, Extended Abstracts and Final Program,
International Symposium on Aerosols in the Mining
and Industrial Work Environment, Minneapolis, 1979.

3. Taylor, C.D., and Jankowski, R.A., The Use of
Instantaneous Samplers to Evaluate the Effectiveness
of Respirable Dust Control Methods in Underground
Mines, Extended Abstracts and Final Program,
International Symposium on Aerosols in the Mining
and Industrial Work Environment, Minneapolis, 1981.

REMARKS   1. The unit comes with a factory calibration attuned to a
representative respirable test dust. A panel-mounted
control allows adjustment of response to match any
reference gravimetric calibration.

NOTES

INHALABLE PARTICULATE MATTER
PERSONAL/PASSIVE/ANALYZER

I-3
PPM, Inc.
Handheld Aerosol
 Monitor
1 of 2

Photograph courtesy of PPM, Inc.

Weight: 1 kg   Dimensions: 31 x 10 x 6 cm

PRINCIPLE OF
OPERATION

Near-forward light scattering. A light-emitting diode coupled to a solid-state photodetector senses the light radiation scattered by micron-sized dust in an open sensing volume.

PERFORMANCE

Lower Detectable Limit: 0.001 mg/m^3

Range: 0 to 2, 0 to 20, or 0 to 200 mg/m^3

Interferences:

Sampling Rate: open sensing volume

Accuracy: within ±25%, typical

Reproducibility:

Zero Drift: ±0.005 mg/m^3/h

Span Drift: 1%/h for battery operation, ±2% for 110 V
             ac, ±10%

OPERATION

Temperature Range: 0 to 40 °C (0 to 55 °C for storage)

Relative Humidity Range: noncondensing

Calibration: referable to gravimetric procedures, factory-
              supplied calibration element

<u>Warmup Time:</u>

<u>Unattended Period:</u> 10 h on 200- and 20-mg range, 5 h on 2-mg
range for battery

<u>Maintenance:</u>

<u>Power:</u> four AA Ni-Cd batteries; rechargeable in unit

FEATURES

<u>Output:</u> LED display; 0- to 2-V analog

<u>Training:</u> none required for sampling

<u>Options:</u> field gravimetric calibration kit

COSTS

HAM sampling kit (includes monitor, instrument case, field
calibration and zero elements, charger, manual, and tools):
$2 500

Field gravimetric calibration kit: $90

MANUFACTURER

PPM, Inc.
11428 Kingston Pike
Knoxville, Tennessee 37922
(615) 966-8796

REFERENCES

A. <u>Specifications:</u>

1. Manufacturer's bulletin 108-5/83.

B. <u>Operations experience:</u>

1. Nagda, N.L., Fortmann, R.C., Koontz, M.D., and
Rector, H.E., Comparison of Instrumentation for
Microenvironmental Monitoring of Respirable
Particulates, <u>Proceedings of the 78th Annual Meeting
of the Air Pollution Control Association</u>, Paper
No. 85-30A.1, Pittsburgh, 1985.

REMARKS

1. The manufacturer is developing a miniature data logger
that is compatible with the HAM system.

2. The HAM can be ordered in an alternative configuration
with a remote sensing head.

INHALABLE PARTICULATE MATTER
STATIONARY/ACTIVE/COLLECTOR

I-4
Sierra-Andersen
Dichotomous Sampler
Series 241
1 of 3

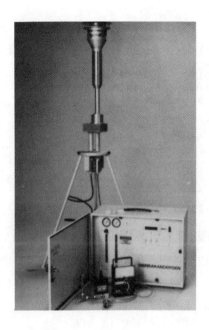

Weight: Control Module, 25 kg
Sampling Module, 7 kg

Dimensions:
Control Module, 41 x 56 x 28 cm
Sampling Module, 162-cm height,
76-cm diameter tripod base bolt
circle; interconnecting tubing,
10 m long

Photograph courtesy of Sierra-Andersen, Inc.

PRINCIPLE OF OPERATION	Size-selective inlet followed by virtual impactor. Ambient air first is accelerated through a nozzle/target impactor to remove particles larger than 10 μm aerodynamic diameter. The sample air containing particles <10 μm then passes through a virtual impactor that has a cut point of 2.5 μm. Fine (<2.5 μm) and coarse (≥2.5 μm) fractions are collected on separate 37-mm TEF-DISC™ Teflon filters. Mass concentration is quantitated gravimetrically.

PERFORMANCE

Lower Detectable Limit:

Range: any ambient particulate concentration

Interferences: the Teflon filters have zero artifact
formation

Sampling Rate: total sample flow: 16.7 L/min
fine fraction: 15 L/min
coarse fraction: 1.67 L/min

I-4
Sierra-Andersen
Dichotomous Sampler
Series 241
2 of 3

Accuracy: constant flow controller: ±5% at 16.7 L/min
            overpressure drop range: 0-35 cm Hg
            standard timer: ±30 min per 7 d
            optional timer: ±2 min per week
            flow meters: ±3% at set flows

Reproducibility: ±5%

Zero Drift:

Span Drift:

OPERATION   Temperature Range: -20 to 40 °C

                 Relative Humidity Range: 0 to 100%

                 Calibration: only flow calibration required for sampling

                 Warmup Time:

                 Unattended Period: defined by sampling schedule

                 Maintenance: routine

                 Power: 110/115 V ac ±10%, 50 to 60 Hz, 6 A max;
                          230 V ac ±10%, 50 Hz, 4 A max (optional)

FEATURES   Output: laboratory report, elapsed timer, flow event
                           circular chart, vacuum gauges

                 Training: recommended

                 Options: digital timer/programmer

COSTS   Series 241 dichotomous sampler: $4 675
                 Digital timer/programmer: $300
                 Model 246-10 field modifications kit (to retrofit
                 15 µm dichotomous samplers): $875

MANUFACTURER   Sierra-Andersen, Inc.
                 4215 Wendell Drive
                 Atlanta, Georgia   30336
                 (404) 691-1910
                 (800) 241-6898

I-4
Sierra-Andersen
Dichotomous Sampler
Series 241
3 of 3

REFERENCES

A. Specifications:

1. Manufacturer's bulletin No. SA-PM10-682.

B. Operations experience:

1. Nagda, N.L., Fortmann, R.C., Koontz, M.D., and Rector, H.E., Comparison of Instrumentation for Microenvironmental Monitoring of Respirable Particulates, Proceedings of the 78th Annual Meeting of the Air Pollution Control Association, Paper No. 85-30A.1, Pittsburgh, 1985.

REMARKS

1. The size-selective inlet has a cut point at 10±1 μm over windspeed of 2-24 km/h; the virtual impactor has a cut point at 2.5 μm; internal losses of the virtual impactor are less than 2% of 0- to 10-μm size range.

2. The manufacturer also offers a field modification kit (model 246-10) to retrofit existing 15-μm samplers for 10-μm performance.

3. If EPA promulgates a 10-μm particulate matter standard, the manufacturer guarantees to obtain EPA reference method approval for this instrument and further guarantees that the model 246-10 inlet to retrofit existing 15-μm samplers will meet EPA performance specifications.

NOTES

INHALABLE PARTICULATE MATTER
STATIONARY/ACTIVE/COLLECTOR

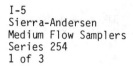

I-5
Sierra-Andersen
Medium Flow Samplers
Series 254
1 of 3

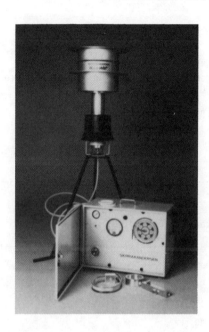

Weight: Control Module, 27 kg
          Sampling Module, 11 kg

Dimensions:
   Control Module, 41 x 56 x 28 cm;
   Sampling Module, 134-cm height;
   Aerosol Inlet, 1.3-m height

Photograph courtesy of Sierra-Andersen, Inc.

PRINCIPLE OF OPERATION	Size selection/filtration. Suspended particles in ambient air enter the 10-μm Med-Flo™ inlet at a flow rate of 6.8 m³/h. The particles are then accelerated through multiple impactor nozzles. By virtue of their larger momentum, particles greater than the 10-μm cut point impact out and are retained in the impaction chamber. The particle fraction smaller than 10 μm is carried vertically upward by the airflow and down the vent tube to the 102-mm Sierra-Andersen TEF-DISC™ Teflon filter where it is uniformly collected.

PERFORMANCE

Lower Detectable Limit:

Range: any ambient particulate concentration

Interferences: Teflon filters have zero artifact formation

Sampling Rate: 11.3 L/min

I-5
Sierra-Andersen
Medium Flow Samplers
Series 254
2 of 3

Accuracy: pneumatic flow controller, ±5% accuracy of
6.8 m^3/h over an inlet pressure drop of
0 to 25 cm Hg; ±10% over an inlet pressure
drop of 0 to 30 cm Hg

Reproducibility: ±3%

Zero Drift:

Span Drift:

OPERATION

Temperature Range: -20 to 40 °C, 600 to 800 mm Hg

Relative Humidity Range: 0 to 100%

Calibration: flow calibration required only for sampling

Warmup Time:

Unattended Period: defined by sampling schedule

Maintenance: routine

Power: 254, 254M: 110/115 V ac, 50 to 60 Hz, 7 A max;
254X, 254MX: 220 V ac, 50 Hz, 4 A max

FEATURES

Output: laboratory report, flow event circular chart,
magnehelic gauge flow indicator

Training: recommended

Options: digital timer/programmer (optional), all functions
digital and quartz crystal controlled; has digital
clock with 1/2-in LED

COSTS

Series 254 Medium Flow Sampler: $3 475
Series 302 Digital Timer/Programmer: $300

MANUFACTURER

Sierra-Andersen, Inc.
4215 Wendell Drive
Atlanta, Georgia 30336
(404) 691-1910
(800) 241-6898

REFERENCES

A. Specifications:

    1. Manufacturer's bulletin, No. SA-PM10-682.

B. Operations experience:

REMARKS

1. The size-selective inlet has a cut point at 10 ± 1 µm over windspeed of 2 to 24 km/h; it meets EPA's expected PM_{10} Federal Reference Method.

2. If EPA promulgates a 10 µm particulate matter standard, the manufacturer guarantees to obtain EPA Reference Method approval for this instrument.

NOTES

INHALABLE PARTICULATE MATTER
PERSONAL/ACTIVE/COLLECTOR

I-6
Sierra Instruments
Marple Personal
 Cascade Impactor
1 of 3

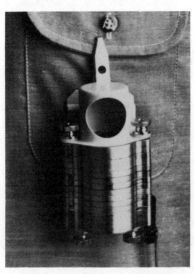

Photograph courtesy of Sierra Instruments, Inc.

Weight: 0.2 kg   Dimensions: 9 x 6 x 4 cm

PRINCIPLE OF
OPERATION

Impaction. Upon entering the inlet, sample air is accelerated through radial slots in the first impaction stage.
Particles larger than the cut point impact on the perforated
collection substrate. The sample airstream then passes to
the next impactor stage, which exhibits a smaller cut point
for impaction, and so on through successively smaller cut
points; remaining fine particles are collected on a backup
filter. The model 294 has four stages, the model 296 has six
stages, and the model 298 has eight stages.

PERFORMANCE

<u>Lower Detectable Limit</u>:

<u>Range</u>: Model 294 (four stages)--cut points at 21, 15, 10,
          3.5 μm
          Model 286 (six stages)--cut points at 10, 6, 3.5,
          1.6, 0.9, 0.5 μm
          Model 296 (eight stages)--cut points at 21, 15, 10,
          6, 3.5, 1.6, 0.9, 0.5 μm

<u>Interferences</u>:

<u>Sampling Rate</u>: 2 L/min

<u>Accuracy</u>:

Reproducibility:

Zero Drift:

Span Drift:

OPERATION   Temperature Range:

Relative Humidity Range:

Calibration:

Warmup Time:

Unattended Period:

Maintenance:

Power:

FEATURES   Output: laboratory report

Training: none required for sampling

Options:

COSTS   294: $775
                     296:   $975
                     298:   $1 175

MANUFACTURER   Sierra Instruments, Inc.
                     P.O. Box 909
                     Carmel Valley, California   93924
                     (408) 659-3177
                     (800) 538-9520

REFERENCES   A.   Specifications:

1.   Manufacturer's bulletin.

B. Operations experience:

1. Nagda, N.L., Fortmann, R.C., Koontz, M.D., and
    Rector, H.E., Comparison of Instrumentation for
    Microenvironmental Monitoring of Respirable
    Particulates, Proceedings of the 78th Annual Meeting
    of the Air Pollution Control Association, Paper
    No. 85-30A.1, Pittsburgh, 1985.

REMARKS

NOTES

INHALABLE PARTICULATE MATTER
PORTABLE/ACTIVE/ANALYZER
STATIONARY/ACTIVE/ANALYZER

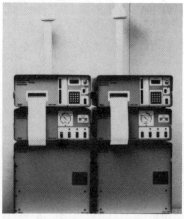

Weight: 4.5 kg (Model 3500); 48 kg (Model 5000)

Dimensions: 31 x 13 x 17 cm (Model 3500)
             38 x 43 x 18 cm, sensor module   ⎫ Model
             38 x 43 x 18 cm, control module ⎬ 5000
             38 x 43 x 42 cm, reservoir module ⎭

Photographs courtesy of TSI, Incorporated

PRINCIPLE OF
OPERATION

Electrostatic precipitation/piezoelectric resonance. The
sample airstream is passed through a cyclone or an impactor
to remove nonrespirable particles (aerodynamic diameter
>3.5 µm). RSP aerosol exiting the impactor is electro-
statically precipitated onto a quartz crystal sensor. The
change in oscillating frequency of the sensing crystal
during the measurement period is proportional to collected
mass. The Model 3500 is battery-powered and portable with
manually initiated sampling periods. The Model 5000 is not
battery powered, but has programmable automatic sampling
cycles for 24-h/d monitoring.

PERFORMANCE

Lower Detectable Limit: approximately 5 µg/m^3 over a
                         10-min averaging time

Range: Model 3500: 0.01 to 10 mg/m^3 (mass); 0.01 to 10 µm
                     (size, 50% cutoff)
        Model 5000: 0.005 to 9.999 mg/m^3 (mass); 0.01 to
                     10 µm (size, 50% cutoff)

Interferences: Changes in relative humidity during a
                single measurement period can cause error.
                Dry, submicrometer, long-chain agglomerated
                particles with no condensed water and no

172   GUIDELINES FOR MONITORING INDOOR AIR QUALITY

I-7
TSI, Incorporated
Piezo Balance
Model 3500 and
 Model 5000
2 of 3

other particles present (e.g., dry diesel
exhaust particles) are not sensed accurately.
Organic vapors are sometimes a positive
interferent.

Sampling Rate: 1 L/min

Accuracy: ±10% or ±0.01 mg/m^3

Reproducibility: ±5%

Zero Drift: automatic rezero at the beginning of every
measurement

Span Drift: crystal sensitivity is an inherent property of
the unbroken piezoelectric quartz crystal; span
does not drift

OPERATION   Temperature Range: 5 to 40 °C

Relative Humidity Range: 10 to 90%

Calibration: internal reference for both collection efficiency and crystal sensitivity

Warmup Time: in a normal room, 5 min or less (the instrument components in contact with the sample
stream must be equilibrated within ±1 °C of
the sample stream temperature)

Unattended Period: up to 4 weeks (Model 5000); Model
3500 is manual

Maintenance: clean sensor crystal after 5-μg accumulation, as indicated by display (Model 3500);
check and refill liquid levels at 1- to 4-week
intervals, refill paper tape (Model 5000);
annual laboratory calibration recommended

Power: rechargeable Ni-Cd, 8-h operation at 50% duty cycle,
15 h recharge needed (Model 3500), single phase ac at
500-W total (Model 5000)

FEATURES   Output: four-digit LED (both); 40-column dot matrix on roll
paper (Model 5000); both analog and digital outputs
compatible with most data systems (Model 5000)

Training: none required for sampling

Options: variety of alternative upper-size cutoffs
           (impactors) ranging from 0.5-10 µm or respirable
           cyclone with 3.5-µm cutoff

COSTS	Model 3500: $4 990
	Model 5000: $16 450

MANUFACTURER   TSI, Incorporated
                      500 Cardigan Road
                      P.O. Box 43394
                      St. Paul, Minnesota   55164
                      (612) 483-0900
                      TELEX: 297-482

REFERENCES   A. Specifications:

                          1. Manufacturer's bulletin, No. TSI 3500/5000-10/80-10M.

                          2. Sem, G., Tsurubayashi, K., and Homma, K., Perfor-
                              mance of the Piezoelectric Microbalance Respirable
                              Aerosol Sensor, Am. Ind. Hyg. Assoc. J., vol. 38,
                              pp. 580-588, 1977.

                          3. Sem, G., and Quant, F., Automatic Piezobalance Respi-
                              rable Aerosol Mass Monitor for Unattended Real-Time
                              Measurements, in Aerosols in the Mining and Indus-
                              trial Work Environment, Vol. 3, Instrumentation,
                              pp. 1039-1054, eds., V. Marple and B. Liu, Ann Arbor
                              Sciences, Ann Arbor, 1982.

                      B. Operations experience:

                          1. Repace, J.L., and Lowrey, A.H., Indoor Pollution,
                              Tobacco Smoke, and Public Health, Science, vol. 208,
                              pp. 464-472, 1980.

                          2. Linch, A.L., Evaluation of Ambient Air Quality by
                              Personal Monitoring, Volume II, Aerosols, Monitor
                              Pumps, Calibration, and Quality Control, CRC Press,
                              Inc., Boca Raton, 1981.

                          3. Quant, F., Nelson, P., and Sem, G., Experimental
                              Measurements of Aerosol Concentrations in Offices,
                              Environ. Int., vol. 8, pp. 223-227, 1982.

REMARKS   1. The Model 3500 supports measurement periods of 24 to
                          120 s (user selectable); the Model 5000 supports
                          measurement periods of 10 s to 2 h.

NOTES

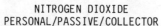

NITROGEN DIOXIDE
PERSONAL/PASSIVE/COLLECTOR

N-1
Air Quality Research,
 Inc., International
Air Check
Nitrogen Dioxide
 Home Test Kit
1 of 3

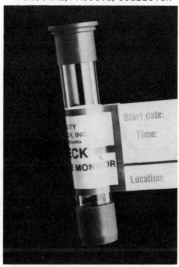

Photograph courtesy of Air Quality Research, Inc., International

Weight: negligible   Dimensions: 7.5-cm length, 1.2-cm diameter

PRINCIPLE OF OPERATION	Molecular diffusion. The sampler consists of three stainless-steel screens coated with triethanolamine (TEA) housed in a plastic tube that is capped when not in use. NO_2 diffuses through the tube at a rate dependent upon Fick's First Law of Diffusion. The TEA-coated screens at the end of the tube maintain a near-zero concentration of NO_2 at the base; therefore, the quantity of NO_2 transferred is related to the ambient concentration and the length of exposure time. Collected NO_2 is quantified spectrophotometrically in a laboratory (Saltzman procedure).
PERFORMANCE	<u>Lower Detectable Limit</u>: 1.5 ppb (three times blank value for 1-week exposure) <u>Range</u>: validated over range of 0 to 30 ppm h <u>Interferences</u>: none <u>Sampling Rate</u>: continuous molecular diffusion <u>Accuracy</u>: <u>Reproducibility</u>: ±15% <u>Zero Drift</u>:

N-1
Air Quality Research,
 Inc., International
Air Check
Nitrogen Dioxide
 Home Test Kit
2 of 3

Span Drift:

OPERATION   Temperature Range: 15 to 30 °C

Relative Humidity Range: noncondensing

Calibration: laboratory standards

Warmup Time:

Unattended Period: 5 to 7 d (recommended minimum exposure
                    for indoor air quality studies)

Maintenance: shelf life has been validated for at least
              6 mo

Power: None

FEATURES   Output: laboratory report

Training: no special training required (see remark 1)

Options:

COSTS   Sampler only: $20 per box of two
                         Sampler plus analysis: $48 per box of two
                         NOTE: these are nominal prices; actual costs depend upon
                                lot sizes

MANUFACTURER   Air Quality Research, Inc., International
                         901 Grayson Street
                         Berkeley, California 94710
                         (415) 644-2097

REFERENCES   A. Specifications:

1. Manufacturer's bulletin.

2. Palmes, E.D., Gunnison, A.F., DiMatto, J., and
    Tomczyk, C., Personal Sampler for Nitrogen Dioxide,
    Am. Ind. Hyg. Assoc. J., vol. 37, pp. 570-577, 1976.

N-1
Air Quality Research,
 Inc., International
Air Check
Nitrogen Dioxide
 Home Test Kit
3 of 3

3. Girman, J.R., Hodgson, A.T., Robisen, B.K., and
    Traynor, G.W., Laboratory Studies of the Temperature
    Dependence of the Palmes NO_2 Passive Sampler,
    Lawrence Berkeley Laboratory, Report No. 16302, 1983.

B. Operations experience:

1. Palmes, E.D., Development and Application of a
    Diffusional Sampler for NO_2, Environ. Int.,
    vol. 6, pp. 97-100, 1981.

2. Leaderer, B.P., Zagraniski, R.T., Berwick, M.,
    Stolwijk, J.A.J., and Qing-Shan, M., Residential
    Exposures to NO_2, SO_2, and HCHO Associated with
    Unvented Kerosene Heaters, Gas Appliances and
    Sidestream Tobacco Smoke, Proceedings of the 3rd
    International Conference on Indoor Air Quality and
    Climate, Stockholm, vol. 4, pp. 151-155, 1984.

REMARKS   1. These devices do not require extensive training for use.
                     However, care must be exercised in proper placement in
                     the field and recording of the exposure interval.

                 2. The sampler, originally developed by Palmes et al. (see
                     reference A.2 above), has been used extensively for
                     indoor air quality testing.

                 3. At temperatures between 5 and 20 °C a 5 to 10% reduction
                     in collection efficiency has been noted due to the
                     solidification of TEA (see reference A.3 above).

                 4. References B.1 and B.2 describe field experience with
                     samplers that are similar to those produced by AQRI.

NOTES

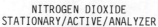

NITROGEN DIOXIDE
STATIONARY/ACTIVE/ANALYZER

N-2
CEA
TGM 555
NO$_2$ Analyzer
1 of 2

Photograph courtesy of CEA Instruments, Inc.

Weight: 14 kg   Dimensions: 51 x 41 x 18 cm

PRINCIPLE OF OPERATION	Automated wet chemistry/colorimetry. Sample air is continuously absorbed in an azo dye-forming reagent. The intensity of the azo dye formed is measured at 550 nm and is directly proportional to the concentration of NO$_2$. Reagent handling and processing is automatic.

PERFORMANCE

Lower Detectable Limit: 0.005 ppm for 0 to 0.15 ppm
full scale

Range: 0 to 0.5 ppm (adjustable from 0 to 0.15 ppm up
to 0 to 10 ppm)

Interferences: none

Sampling Rate: 250 mL/min, continuous

Accuracy:

Reproducibility: 1%

Zero Drift: <2% (72 h)

Span Drift: <2% (72 h)

OPERATION

Temperature Range: 15 to 25 °C

Relative Humidity Range: 5 to 95%

N-2
CEA
TGM 555
NO$_2$ Analyzer
2 of 2

Calibration: with liquid standards, permeation tubes, or
              gas-phase titration

Warmup Time: 20 min

Unattended Period: 18 h on fully charged batteries

Maintenance: tubing in the peristaltic pump should be
              changed once a month

Power: 12 V dc unregulated, 4 W 115/230 V ac, 50/60 Hz

FEATURES

Output: digital panel meter
         0 to 1 V at 0 to 2.0 milliamps recorder output

Training: none required for sampling

Options: Reaction Chamber (for converting NO to NO$_2$)
          Stream Splitter (to extend range by a factor
          of 10 or 100)

COSTS

TGM 555: $5 340
Reaction Chamber: $150
Stream Splitter: $295

MANUFACTURER

CEA Instruments, Inc.
16 Chestnut Street
P.O. Box 303
Emerson, New Jersey 07630
(201) 967-5660

REFERENCES

A. Specifications:

    1. Manufacturer's bulletin.

B. Operations experience:

REMARKS

1. The unit can be fitted for monitoring NO$_x$ by installing
    a solid oxidant converter, which converts NO to NO$_2$.

NITROGEN DIOXIDE
PORTABLE/ACTIVE/ANALYZER

N-3
CSI
Model 2200
Portable NO$_x$
 Analyzer
1 of 3

Photograph courtesy of Columbia Scientific Industries Corporation

Weight: 9 kg   Dimensions: 20 x 18 x 46 cm

PRINCIPLE OF OPERATION	Chemiluminescence. Sample air is initially routed to a reaction chamber where chemiluminescent reactions with O$_3$ are detected and quantified by a photomultiplier tube, producing the NO signal, which is stored electronically. A second air sample is routed through an NO$_2$-to-NO converter and then to the O$_3$ reaction chamber, producing the NO$_x$ signal. The NO$_2$ value is electronically calculated by subtracting NO from NO$_x$.

PERFORMANCE

Lower Detectable Limit: 0.020 ppm (5-s time constant set-
                         ting); 0.010 ppm (60-s time con-
                         stant setting)

Range: 0.5, 1.0, 2.0, or 5.0 ppm

Interferences: total interference equivalent for H$_2$O, SO$_2$,
                NO, and NH$_3$ is 0.010 ppm on the NO$_x$ channel

Sampling Rate: 700 mL/min, continuous

Accuracy: depends on calibration source accuracy

Reproducibility: 2% of full scale

Zero Drift: ±0.005 ppm for 12 h
             (±0.0005 ppm/°C at 15 to 35 °C)

Span Drift: ±2% for 12 h
             (±0.3%/°C at 15 to 35 °C)

N-3
CSI
Model 2200
Portable NO_x
 Analyzer
2 of 3

OPERATION	Temperature Range: 10 to 40 °C
	Relative Humidity Range: 5 to 95%
	Calibration: gas-phase titration
	Warmup Time: 30 min
	Unattended Period: 2 h on internal battery, 5 h with external battery pack, 7 or more days with ac adapter/charger
	Maintenance: converter life is normally 1 yr; operating manual describes routine maintenance
	Power: 12 V dc
FEATURES	Output: 0 to 1 V dc for NO, NO_x, NO_2; 12 V dc for optional battery-operated chart recorder, 10 V dc, 1-mA alarm output plus analog panel meter
	Training: recommended
	Options: portable recorder, 1-in/h chart speed, 12-V auto lighter cable assembly; auxiliary battery pack (provides up to 8 h of additional battery operation)
COSTS	Model 2200: $7 350 Recorder: $725 12-V Auto Lighter Cable: $98 Auxiliary Battery Pack: $575
MANUFACTURER	Columbia Scientific Industries Corporation P.O. Box 9908 Austin, Texas 78766 (512) 258-5191 (800) 531-5003 TWX: 910-874-1364
REFERENCES	A. Specifications: 1. Manufacturer's bulletin. B. Operations experience:

N-3
CSI
Model 2200
Portable NO_x
 Analyzer
3 of 3

REMARKS

1. Unit offers automatic failure diagnosis and display system.

2. Photomultiplier tube temperature is maintained at 20 °C by a thermoelectric cooling system to minimize noise and zero drift.

3. Gas reaction chamber is regulated at 42 °C to minimize span errors.

4. CSI offers training sessions in Austin, Texas, on a monthly basis.

NOTES

NITROGEN DIOXIDE
PERSONAL/PASSIVE/COLLECTOR

N-4
Du Pont
PRO-TEK™
NO_2 Passive Dosimeter
Type C30
1 of 3

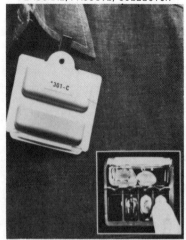

Photograph courtesy of Du Pont Applied Technology Instruments

Weight: negligible   Dimensions: 7.6 x 7.1 x 0.89 cm

PRINCIPLE OF OPERATION

Molecular diffusion/sorption. Collection relies upon molecular diffusion to deliver sample air to a liquid sorbent solution at a constant rate. After exposure, the sorbent is analyzed in a laboratory spectrophotometer or PRO-TEK™ PT-3 Readout for NO_2 content and subsequently the time-weighted average concentration. Laboratory validation has been conducted only for up to 8-h exposures.

PERFORMANCE

Lower Detectable Limit: 10 ppm h, when analyzed in a PT-3 readout; 1.5 ppm h, when analyzed in a laboratory spectrophotometer

Range: 10 to 100 ppm h (PT-3 readout)
1.5 to 200 ppm h (laboratory spectrophotometer)

Interferences: The major known interferences are SO_2, nitrates, O_3, and strong oxidizing agents. These interferences will not affect the dosimeter for NO_2 because they have little or no affinity for the absorbing solution.

Sampling Rate: diffusion, continuous

Accuracy: ±18.2% (overall accuracy)

Reproducibility: ±7.5%

Zero Drift:

N-4
Du Pont
PRO-TEK™
NO_2 Passive Dosimeter
Type C30
2 of 3

	Span Drift:

OPERATION	Temperature Range: 4 to 49 °C
	Relative Humidity Range:
	Calibration: laboratory standards
	Warmup Time:
	Unattended Period: 8 h (longer periods are possible)
	Maintenance: shelf life of unexposed samplers under refrigeration is 6 mo; exposed samplers have a refrigerated shelf life of 3 weeks; unrefrigerated shelf life of exposed samplers is 2 weeks
	Power: none required for sampling

FEATURES	Output: laboratory report
	Training: none required for sampling
	Options:

COSTS	Type C30, 10 per box:

                  1-10 boxes: $259 (Order Code 5115)
           11-25 boxes: $233 (Order Code 5116)
             26+ boxes: $207 (Order Code 5117)

MANUFACTURER	Du Pont Applied Technology Instruments Box 10 Kennett Square, Pennsylvania 19348 (215) 444-4493 (800) 344-4900

REFERENCES	A. Specifications:
	1. Manufacturer's analysis instructions.

2. Kring, E.V., Lautenberger, W.J., Baker, W.B.,
    Douglas, J.J., and Hoffman, R.A., A New Passive
    Colorimetric Air Monitoring Badge System for Ammonia,
    Sulfur Dioxide, and Nitrogen Dioxide, Am. Ind.
    Hyg. Assoc. J., vol. 42, pp. 373-381, 1981.

B. Operations experience:

1. Woebkenberg, M.L., A Comparison of Three Passive
    Personal Sampling Methods for NO$_2$, Am. Ind. Hyg.
    Assoc. J., vol. 43, pp. 553-561, 1982.

2. Laboratory Validation Report, PRO-TEK™ Nitrogen
    Dioxide Badge, Type C30, Du Pont (2/2/81).

REMARKS	1.	Immediate readout is possible using PT-3 colorimeter because chemical reagents are stored inside the badge. However, greater sensitivity is possible from a laboratory spectrophotometer.

NOTES

NITROGEN DIOXIDE
PORTABLE/ACTIVE/ANALYZER

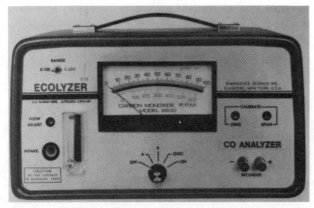

Photograph courtesy of Energetics Science, Inc.
(CO analyzer shown is identical in appearance to NO$_2$ analyzer)

Weight: 4.5 kg   Dimensions: 17.8 x 17.8 x 33 cm

PRINCIPLE OF OPERATION	Electrochemistry. Ambient air is drawn into an electro-chemical sensor, producing a signal that is proportional to NO$_2$ concentrations.

PERFORMANCE

Lower Detectable Limit: <0.5 ppm

Range: 0 to 2 ppm, 0 to 10 ppm

Interferences: expressed as ppm of interferent needed to give 1 ppm deflection (testing performed with Purafil filter installed): C$_2$H$_2$ (acetylene) = 2, C$_3$H$_8$ (propane) = 20 000, CO = 1 000, H$_2$ = 3 000

Sampling Rate: 700 mL/min, continuous

Accuracy: ±1%

Reproducibility: ±1%

Zero Drift: <±0.5 ppm/d

Span Drift: 1%/d

N-5
Energetics Science
2000 Series
NO_2 Analyzer
2 of 2

OPERATION	Temperature Range: 0 to 40 °C
	Relative Humidity Range: unaffected by water vapor
	Calibration: standard gas mixture
	Warmup Time: specified only as "brief"
	Unattended Period: 8+ h on battery
	Maintenance: 1-yr sensor warranty; boards replaceable in field
	Power: 105-125 V ac at 50-60 Hz; Ni-Cd batteries with built-in recharger
FEATURES	Output: panel meter with parallax mirror; 0- to 1-V dc recorder output
	Training: none required for sampling
	Options: dc-powered recorder ac-powered recorder
COSTS	Model 2000 NO_2 monitor: $1 900 dc recorder: $550 ac recorder: $450
MANUFACTURER	Energetics Science, Inc. 6 Skyline Drive Hawthorne, New York 10532 (914) 592-3010
REFERENCES	A. Specifications: 1. Manufacturer's bulletin. B. Operations experience:
REMARKS	1. Another version of the instrument (7000 series) allows any two of the following to be paired in the same chassis: CO, NO_2, NO, and H_2S.
	2. These units were originally designed for workplace monitoring (TLV = 3 ppm) and may be of limited use for nonindustrial environments unless high concentrations are present.

NITROGEN DIOXIDE
PORTABLE/ACTIVE/ANALYZER

Photograph courtesy of Interscan Corporation

Weight: 3.6 kg (1150)   Dimensions: 18 x 15 x 29 cm (1150)
          20.0 kg (4150)   18 x 10 x 23 cm (4150)

PRINCIPLE OF   Electrochemistry. Gas molecules from the moving sample
OPERATION   airstream pass through a diffusion medium and are adsorbed
                    onto an electrocatalytic sensing electrode where subsequent
                    reactions generate an electric current. The diffusion-
                    limited current is linearly proportional to NO_2
                    concentration.

PERFORMANCE   Lower Detectable Limit: 1% of full scale

                    Range: 0 to 2 ppm, 0 to 10 ppm

                    Interferences: expressed as ppm of interferent needed to
                                     give 1 ppm deflection: NH_3 = 35. Gas
                                     scrubbers are needed to remove Cl_2,
                                     ethyl mercaptans, methyl mercaptans,
                                     SO_2, and H_2S when these interferents are
                                     at concentrations equivalent to NO_2.

                    Sampling Rate: continuous

                    Accuracy: ±2% of full scale

                    Reproducibility: ±0.5% of full scale

                    Zero Drift: ±1% full scale in 24 h

                    Span Drift: <±2% full scale in 24 h

N-6
Interscan
Models 1150 and
 4150
NO$_2$ Analyzers
2 of 2

OPERATION   Temperature Range: 10 to 120 °F

                          Relative Humidity Range:

                          Calibration: standard gas mixture

                          Warmup Time:

                          Unattended Period: 10 h on battery power

                          Maintenance:

                          Power: four alkaline MnO$_2$ batteries for amplifier, two
                                  Ni-Cd for pumps LCD; one HgO battery for bias
                                  amplifier reference

FEATURES   Output: 0 to 100 mV full scale

                          Training: none required for sampling

                          Options: 1150, audible and visual alarm;
                                    4150, audible alarm

COSTS   Model 1150: $1 675
                          Model 4150: $1 895

MANUFACTURER   Interscan Corporation
                          P.O. Box 2496
                          21700 Nordhoff Street
                          Chatsworth, California 91311
                          (213) 882-2331
                          TELEX: 67-4897

REFERENCES   A. Specifications:

                          1. Manufacturer's bulletin.

                          B. Operations experience:

REMARKS   1. These units were originally designed for workplace
                              monitoring (TLV = 3 ppm) and may be of limited use for
                              nonindustrial environments unless high concentrations
                              are present.

<table>
<tr><td></td><td>N-7</td></tr>
</table>

NITROGEN DIOXIDE
PERSONAL/PASSIVE/ANALYZER

N-7
Interscan
Model 5150
NO_2 Analyzer and
CX-5 Data Interface
1 of 3

Photograph courtesy of Interscan Corporation
(Model 5140 shown is identical in appearance to Model 5150)

Weight: 0.7 kg (5150)   Dimensions: 15 x 8 x 5 cm (5150)
         0.6 kg (CX-5)   7 x 16 x 3 cm (CX-5)

PRINCIPLE OF   Diffusion/electrochemistry. NO_2 diffuses into an electro-
OPERATION   chemical cell, producing a signal proportional to NO_2 con-
                   centrations. The signal is digitized, incorporated into
                   1-min averages, and stored. Nondestructive recovery of each
                   1-min average is accomplished through a separate data reader.
                   Data storage capacity is 2 048 1-min averages. Stored data
                   can be retrieved through the CX-5 interface and transferred
                   to a computer for further processing.

PERFORMANCE   Lower Detectable Limit: 0.5% of full scale

                   Range: 0 to 30 ppm

                   Interferences: expressed as ppm of interferent needed to
                                   give 1 ppm deflection: NH_3 = 35. Special
                                   filters are required to remove Cl_2,
                                   ethyl mercaptans, methyl mercaptans, and
                                   SO_2 when these interferents are at
                                   concentrations equivalent to NO_2.

                   Sampling Rate: diffusion, continuous

                   Accuracy: ±2% of reading, ±1 least significant digit (LSD),
                              ±0.5% of full scale (digital)

N-7
Interscan
Model 5150
NO$_2$ Analyzer and
 CX-5 Data Interface
2 of 3

	Reproducibility: ±1% reading, ±1 LSD
	Zero Drift: ±1% reading, ±1 LSD in 24 h
	Span Drift: ±1% reading, ±1 LSD in 24 h
OPERATION	Temperature Range: 30 to 120 °F
	Relative Humidity Range: 1 to 100%
	Calibration: standard gas mixture
	Warmup Time: <5 min
	Unattended Period: up to 34 h
	Maintenance: calibration, battery replacement, sensor replacement
	Power: long-life 9-V battery (alkaline MnO$_2$, NEDA type 1604A); battery life is 125 h continuous operation
FEATURES	Output: printout from data reader (see remark 1)
	Training: none required for sampling
	Options:
COSTS	5140: $1 145
	CX-5: $1 375
MANUFACTURER	Interscan Corporation
	P.O. Box 2496
	21700 Nordhoff Street
	Chatsworth, California 91311
	(213) 882-2331
	TELEX: 67-4897
REFERENCES	A. Specifications:
	1. Manufacturer's bulletin.
	B. Operations experience:

N-7
Interscan
Model 5150
NO_2 Analyzer and
  CX-5 Data Interface
3 of 3

REMARKS

1. The CX-5 interface allows nondestructive retrieval of all 1-min averages stored in the dosimeter memory. The interface is programmed to give time-selected and time-weighted averages, maximum short-term exposure levels, and the time at which they occur. Peak concentration and the time at which it occurs is also given. One CX-5 can service many 5000 series dosimeters.

2. Data readout may also be accomplished by a device available from:

   Metrosonics, Inc.
   P.O. Box 23075
   Rochester, New York 14692
   (716) 334-7300

3. These units were originally designed for workplace monitoring (TLV = 3 ppm), and may be of limited use for nonindustrial environments unless high concentrations are present.

NOTES

NITROGEN DIOXIDE
PERSONAL/PASSIVE/COLLECTOR

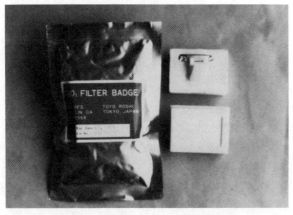

Illustration courtesy of Micro Filtration Systems

Weight: negligible   Dimensions: 5 x 4 x 1 cm

PRINCIPLE OF   Molecular diffusion/sorption. A filter treated with tri-
OPERATION   ethanolamine (TEA) adsorbs NO_2 that diffuses through a
                  series of hydrophobic fiber filters that suppress wind
                  effects. Sorbed NO_2 is quantitated spectrophotometrically.

PERFORMANCE   Lower Detectable Limit: 66 ppb h

                  Range: Up to 10^6 ppb h, theoretical

                  Interferences: adsorption rate for NO_2 may vary by as
                                  much as 20% under wind velocities between
                                  0.15 and 4.0 m/s. The effect of relative
                                  humidity (between 40 and 80% RH) is less
                                  than that of wind velocity. Maximum adsorp-
                                  tion rate occurs at 60% RH.

                  Sampling Rate: diffusion, continuous

                  Accuracy: ±20%

                  Reproducibility: <4.8%

                  Zero Drift:

                  Span Drift:

OPERATION   Temperature Range: room temperature

                  Relative Humidity Range: 40 to 90%

Calibration:　standard curve for laboratory analysis
　　　　　　　　constructed by user from known standards

Warmup Time:

Unattended Period:　24 h to 1 week; up to 1 mo

Maintenance:

Power:　none required for sampling

FEATURES　　　　　　Output:　laboratory report

Training:　none required for sampling

Options:

COSTS　　　　　　　NO$_2$ filter badge:　$11.65 each

MANUFACTURER　　　Micro Filtration Systems
　　　　　　　　　6800 Sierra Court
　　　　　　　　　Dublin, California　94566
　　　　　　　　　(415) 828-6010

REFERENCES　　　　A.　Specifications:

　　　　　　　　　1.　Yanagisawa, Y., and Nishimura, H., A Badge-Type
　　　　　　　　　　　Personal Sampler for NO$_2$ to be Used in the
　　　　　　　　　　　Living Environment, Presented at the Fifth Clean
　　　　　　　　　　　Air Congress, Buenos Aires, 1980.

　　　　　　　　　2.　Yanagisawa, Y., and Nishimura, H., Badge-Type
　　　　　　　　　　　Personal Sampler for Measurement of Personal Expo-
　　　　　　　　　　　sure to NO$_2$ and NO in Ambient Air, Presented
　　　　　　　　　　　at the International Symposium on Indoor Air
　　　　　　　　　　　Pollution, Health and Energy Conservation, Amherst,
　　　　　　　　　　　1981.

　　　　　　　　　B.　Operations experience:

　　　　　　　　　1.　Both references above summarize specifications as
　　　　　　　　　　　well as field use.

N-8
Toyo Roshi
NO$_2$ Badge
3 of 3

REMARKS

1. The badge can be converted for collecting NO$_x$ (NO + NO$_2$) by treating intervening filters with a 5% chromium tri- oxide solution to oxidize NO to NO$_2$ as it diffuses to the sorbent filter. NO$_2$ diffuses through unaltered and is adsorbed.

2. Laboratory analysis is spectrophotometric and uses easily obtained reagents.

NOTES

OZONE
PORTABLE/ACTIVE/ANALYZER

Photograph courtesy of Columbia Scientific Industries Corporation (CSI)
(Model 2200 shown is identical in appearance to CSI 2000)

Weight: 7.7 kg (9.9 kg with   Dimensions: 20 x 18 x 46 cm
          optional battery pack)

PRINCIPLE OF OPERATION	Chemiluminescence. Photometric detection of the flameless reaction of ethylene gas with O_3.

PERFORMANCE	Lower Detectable Limit: 0.004 ppm (on 5-s filter setting) 0.001 ppm (on 60-s filter setting) Range: 0 to 0.10, 0 to 0.20, 0 to 0.50, and 0 to 1.00 ppm Interferences: <0.06 ppm total for H_2O, CO, and H_2S Sampling Rate: 700 mL/min, continuous Accuracy: Reproducibility: ±1.0% of full scale Zero Drift: ±0.002 ppm for 12 h (±0.0002 ppm/°C at 10 to 35 °C) Span Drift: ±1% for 12 h (±2%/°C at 10 to 35 °C)

OPERATION	Temperature Range: 10 to 40 °C Relative Humidity Range: 5 to 95% Calibration: gas-phase titration

O-1
CSI
Model 2000
Portable Ozone Meter
2 of 2

Warmup Time: 30 min

Unattended Period: 8 h for battery operation

Maintenance:

Power: 14 V dc (also 120 or 230 V ac with charger/adapter)

FEATURES   Output: panel meter, 0 to 1.0 or 0.100 mV recorder output

Training: recommended

Options: battery pack
          battery charger

COSTS $6 750 (includes battery and charger)

MANUFACTURER   Columbia Scientific Industries Corporation
                     P.O. Box 9908
                     Austin, Texas
                     (512) 258-5191
                     (800) 531-5003

REFERENCES   A. Specifications:

                     1. Manufacturer's bulletin.

                     B. Operations experience:

REMARKS   1. The Model 2000 is an EPA-Designated Reference Method
                         for ozone.

RADON
STATIONARY/PASSIVE/COLLECTOR

R-1
AeroVironment
PRM LR-5
1 of 3

Photograph courtesy of AeroVironment, Inc.

Weight: 9 kg

Dimensions: 51-cm high x
23-cm diameter

PRINCIPLE OF
OPERATION

Electrostatic collection/thermoluminescence dosimetry.
Ambient Rn diffuses into a sensitive chamber where subsequent disintegrations of ions are electrostatically
focused onto a thermoluminescent dosimeter (TLD) chip held
at negative potential in a 900- to 1200-V electrostatic
field. Each alpha particle striking the chip creates
metastable defects in the crystal, which can be read and
related to integrated Rn concentration. A water-impermeable
membrane keeps the chamber dry while allowing Rn to diffuse
in through the bottom. With the membrane, a desiccant
material is not needed, greatly extending the possible
sampling times in humid climates, and eliminating desiccant
drying. A second TLD chip is exposed away from the
electrostatic field (at the base of the housing) to check
background levels of gamma radiation.

PERFORMANCE

Lower Detectable Limit: 0.03 pCi/L/week (laboratory
conditions), 0.2 pCi/L/week
(adverse field conditions)

Range: 0.03 pCi/L to 10^4 pCi/L (1-week exposure)

Interferences:

Sampling Rate: diffusion, continuous

Accuracy:

R-1
AeroVironment
PRM LR-5
2 of 3

                        Reproducibility:

                        Zero Drift:

                        Span Drift:

OPERATION   Temperature Range: -45 to 65 °C

                        Relative Humidity Range: 0 to 100% for extended periods

                        Calibration: laboratory calibration available

                        Warmup Time: none

                        Unattended Period: <1 week to 12 mo

                        Maintenance: battery voltage check

                        Power: four Eveready Mini-max No. 493 batteries

---

FEATURES   Output: counts from TLD reader

                        Training: none required for sampling

                        Options: spare TLD holders and replacement batteries

---

COSTS $595 each (complete with batteries, good for 1 yr); two TLD
                        chip holders (more available on request); quantity discount
                        available

MANUFACTURER   AeroVironment, Inc.
                        5680 South Syracuse Circle #300
                        Englewood, Colorado 80111
                        (303) 771-3586

                        Head Office:
                        145 Vista Avenue
                        Pasadena, California 91107
                        (213) 449-4392

REFERENCES   A. Specifications:

                            1. Manufacturer's bulletin.

R-1
AeroVironment
PRM LR-5
3 of 3

    2.   George, A.C., A Passive Environmental Radon Monitor,
        in Radon Workshop, ed. Breslin, A.J., U.S. Energy
        Research and Development Administration, Report
        HASL-325, Health and Safety Laboratory, New York,
        pp. 25-30, 1977.

B. Operations experience:

    1.   Friedland, S.S., Rathbun, L., and Goldstein, A.M.,
        Radon Monitoring: Uranium Mill Field Experience
        with a Passive Detector, AeroVironment, Inc.,
        Pasadena, 1980.

REMARKS

1. This instrument is based on the Passive Environmental
Radon Monitor (PERM) developed at the DOE Environmental
Measurements Laboratory (George 1977) and modified
by AeroVironment (Friedland, Rathbun, and Goldstein
1980).

2. However, nondesiccant membrane allows exposure period
of 1 yr or more, limited only by battery life.

NOTES

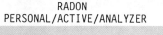

RADON
PERSONAL/ACTIVE/ANALYZER

R-2
Alpha Nuclear
500 Series
Alpha PRISM System
1 of 3

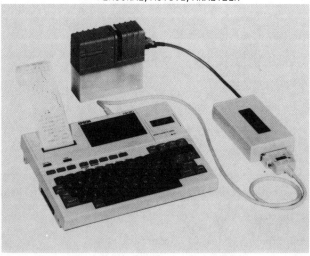

Photograph courtesy of Alpha Nuclear

Weight: 1.6 kg   Dimensions: 15 x 15 x 17 cm

PRINCIPLE OF OPERATION	Filtration/gross alpha counting. Rn progeny are collected on a filter. Subsequent alpha activity is measured by a diffused junction detector permanently facing the sample deposit. Resulting electronic pulses are stored in microprocessor memory. Data are retrieved through an RS-232 interface to provide for computerized data reduction.

PERFORMANCE

Lower Detectable Limit: approximately 0.001 WL at 100 mL/min flow rate

Range: <0.001 to >40 WL

Interferences:

Sampling Rate: 50, 100, or 200 mL/min, selectable

Accuracy:

Reproducibility:

Zero Drift:

Span Drift:

OPERATION

Temperature Range: -20 to 50 °C

Relative Humidity Range:

R-2
Alpha Nuclear
500 Series
Alpha PRISM System
2 of 3

Calibration: standard alpha source

Warmup Time:

Unattended Period: up to 24 h on battery

Maintenance: 12 V rechargeable battery

Power:

FEATURES

Output: formatted summary on paper tape; microcassette

Training: none required for sampling

Options: Model 540 interface
          Model 520 readout system
          Model 505 calibration alpha source
          Model 551A battery pack (12 h)
          Model 551B battery pack (24 h)
          Model 552 battery charger

COSTS

Model 550 air sampling module (includes flow regulated pump,
filter receptacle, alpha detector and microprocessor elec-
tronics--requires battery): $2 200

Model 540 Interface (RS-232 output): $800

Model 520 Readout System (Epson XS-20 with 32K memory,
printer, LCD display, and cassette): $1 100

Model 505 Calibration Source: $200

Model 551A Battery Pack (12 h): $180

Model 551B Battery Pack (24 h): $225

Model 552 Battery Charger: $60

Model 501 Alpha Filter Cards (package of 100): $75

MANUFACTURER

Alpha Nuclear
6380 B Viscount Road
Mississauga, Ontario, Canada L4V 1H3
(416) 676-1364
TELEX: 06-983611

R-2
Alpha Nuclear
500 Series
Alpha PRISM System
3 of 3

REFERENCES   A. Specifications:

                    1. Manufacturer's bulletin, Technical Data 981.

                    B. Operations experience:

REMARKS   1. One readout system can serve a number of dosimeters.

                    2. The Model 540 interface can be programmed to acquire
                        data for incremental time intervals. In this mode the
                        interface has to remain connected to the dosimeter dur-
                        ing the measurement. At the end of sampling, the read-
                        out system is reconnected and data is transferred on an
                        incremental basis to provide a time series.

NOTES

RADON
STATIONARY/ACTIVE/ANALYZER

R-3
Eberline
RGM-2
Radon Gas Monitor
1 of 2

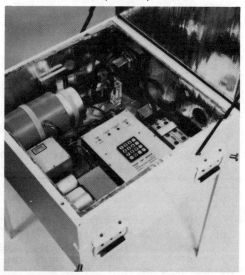

Photograph courtesy of Eberline Instrument Corporation

Weight: 45 kg   Dimensions: 66 x 71 x 31 cm
                                         (sits on legs that are 91 cm)

PRINCIPLE OF   Alpha scintillation. Sample air is filtered and drawn
OPERATION   at a constant rate through a zinc sulfide-coated scin-
                    tillation cell. The scintillation cell is optically
                    coupled to a photomultiplier. Resulting alpha activity
                    is proportional to average Rn gas concentration.

PERFORMANCE   Lower Detectable Limit: 0.3 pCi/L

                    Range: >1 000 pCi/L

                    Interferences: Rn^{220}

                    Sampling Rate:

                    Accuracy: near ±10% at >1 pCi/L for 1 h

                    Reproducibility:

                    Zero Drift:

                    Span Drift:

OPERATION   Temperature Range: -20 to 43 °C

R-3
Eberline
RGM-2
Radon Gas Monitor
2 of 2

Relative Humidity Range:

Calibration:

Warmup Time:

Unattended Period:

Maintenance: changing dust filter on sample intake,
               changing dust filter on cooling air intake,
               replacing printer paper

Power: 115/230 V ac at approximately 1 A

FEATURES   Output: alpha numeric printer, 21 characters per line,
                              electric writing

                    Training: none required for sampling

                    Options:

COSTS $8 800

MANUFACTURER   Eberline Instrument Corporation
                    P.O. Box 2108
                    Santa Fe, New Mexico   87501
                    (505) 471-3232
                    TWX: 910-985-0678

REFERENCES   A.   Specifications:

                         1.   Manufacturer's brochure.

                         2.   Marley, M., and Geiger, E., Continuous Radon Progeny
                              and Gas Monitor, Presented at the International
                              Symposium on Radiation Protection for Uranium Mining
                              and Milling, Albuquerque, 1977.

                    B.   Operations experience:

REMARKS   1.   There is a lag between changes in Rn concentration and
                         corresponding instrument response because the buildup of
                         alpha emitters (Rn progeny formed in the cell that
                         attach to the walls) does not reach equilibrium immedi-
                         ately. Although the correspondence between hourly Rn
                         concentration and instrument output is affected by this
                         lag, the daily average concentration is correct within
                         a few percent.

RADON
PORTABLE/ACTIVE/ANALYZER

R-4
Eberline
Working Level
 Monitor
1 of 3

Photograph courtesy of Eberline Instrument Corporation

Weight: 2.6 kg (WLM-1)   Dimensions: 14.6 x 11.7 x 20 cm (WLM-1)
          6.8 kg (WLR-1)   35.6 x 45.7 x 15.2 cm (WLR-1)

PRINCIPLE OF
OPERATION

Filtration/gross alpha counting. Rn progeny are collected
on a filter and consequent alpha activity is measured using
a silicon-diffused junction detector. A microprocessor
counts and stores detected alpha pulses. The microprocessor
also controls the sampling pump and records decay (tail)
measurements after the sampling interval is terminated.
Length of sample interval and detail of tail data are
operator-selectable by key-pad entries on readout unit.
Data are retrieved through a separate readout unit that also
calculates working levels with percent thoron daughters.

PERFORMANCE

Lower Detectable Limit: 2×10^{-5} WL (99% confidence level
                         based on background of 0.1 counts
                         per min and 168 h sample time)

Range: capable of measuring naturally occurring back-
        ground levels with an upper limit as indicated
        below:

        Based on 200 1-min intervals, 1.5×10^3 WL
        Based on 168 1-h intervals, 1×10^2 WL

Interferences: cosmic radiation, long-lived alpha emitters
                such as uranium and thorium

Sampling Rate: 0.12 to 0.18 L/min, continuous; intervals
                are selectable

R-4
Eberline
Working Level
 Monitor
2 of 3

	Accuracy: <5% maximum error under cases of extreme disequilibrium, plus any error induced by calibration. Typical accuracy is ±5%
	Reproducibility:
	Zero Drift: none
	Span Drift: none
OPERATION	Temperature Range: 30 to 120 °F
	Relative Humidity Range: 0 to 90% noncondensing
	Calibration: americium or thorium alpha source along with flow rate calibration
	Warmup Time: <1 min
	Unattended Period: 168-h data run plus 4-h tail acquisition with extended standby
	Maintenance: exchange sample filter; recharge battery
	Power: 6-V gel cell for sampler (6 A h); ac power for readout
FEATURES	Output: (readout unit) electrosensitive printer, 21 characters per line, two lines/s, operator interactive alpha/numeric LCD display
	Training: none required for sampling
	Options: battery charging stations
COSTS	Approximately $2 000 each, WLM-1; $3 000 each, WLR-1
MANUFACTURER	Eberline Instrument Corporation P.O. Box 2108 Santa Fe, New Mexico 87501 (505) 471-3232 TWX: 910-985-0678

REFERENCES   A. Specifications:

                    1. Beard, R., Geiger, E.L., Waligora, S.J., Fraim, F.,
                        Gunther, B., and Tafreshi, A., Eberline's New
                        Microcomputer Based Radon Daughter Instrument,
                        Presented at the International Symposium on Indoor
                        Air Pollution, Health and Energy Conservation,
                        Amherst, 1981.

                    B. Operations experience:

                    1. Prototypes have been tested at U.S. Bureau of Mines,
                        Denver, Colorado.

REMARKS   1. The readout unit (Model WLR-1) is required to service
                        the monitor (Model WLM-1). One readout unit can service
                        many WLM-1s.

NOTES

<table>
<tr><td>RADON/RADON PROGENY
PORTABLE/ACTIVE/ANALYZER</td><td>R-5
EDA
RDA-200
Radon/Radon Daughter
 Detector
1 of 3</td></tr>
</table>

Photograph courtesy of EDA Instruments, Inc.

Dimensions: console, 12.7 x 16.5 x 20 cm
total system packaged, 61 x 61
x 35.5 cm; gas cell, 5.3 cm
diameter x 7.3 cm

Weight: console, 1.7 kg
system, 8.0 kg

PRINCIPLE OF OPERATION	Alpha scintillation. A known volume of sample air is drawn through a sampling train composed of a filter at the inlet followed by a gas scintillation cell and a user-supplied pump. Rn daughter products collect in the filter; the gas cell retains a sample of Rn in air. The filter is placed in a scintillation tray for counting in the detector; the gas cell is placed directly into the detector for counting. Details of sampling (flow rate, duration) and subsequent alpha counting (time factors) are determined by the operator and the technique employed.

PERFORMANCE

Lower Detectable Limit: dictated by background and
technique

Working Level: 0.03 Kusnetz
0.01 Tsivolglou
0.01 Rolle

Radon: below 1 pCi/L

Range: 0-99, 999 counts (up to 15 000 cpm without loss of
sensitivity); counting periods of 1, 2, 5, 10,
and 60 min selectable plus manual

Interferences:

R-5
EDA
RDA-200
Radon/Radon Daughter
 Detector
2 of 3

Sampling Rate: specified by user and technique

Accuracy:

Reproducibility:

Zero Drift:

Span Drift:

OPERATION   Temperature Range: -30 to 40 °C

Relative Humidity Range:

Calibration: Rn-222 standard gas source and americium-241
                          disc

Warmup Time:

Unattended Period:

Maintenance:

Power: eight C cells standard; external battery pack or ac
             line source optional

FEATURES   Output: five-digit LED

Training: recommended

Options: RDU-200--degassing system for determinations
                          from water and sediments
              RDX-207--americium-241 calibration disc
              RDX-261--battery charger
              RDX-263--external ac/dc converter
              RDX-251--end of counting audio alarm
              (Various air pumps, flowmeters, and specialized
              detector cells also available)

COSTS   RDA-200: $4 450 (includes detector console, radium test
                 cell, five double swagelock Rn gas cells, five scintillator
                 trays, two filter holders, 100 0.8 µm filters, eight C cell
                 batteries, and manual)
                 RDU-200: $1 950
                 RDX-207: $ 650
                 RDX-261: $ 250
                 RDX-263: $ 185
                 RDX-251: $ 75

R-5
EDA
RDA-200
Radon/Radon Daughter
 Detector
3 of 3

MANUFACTURER	EDA Instruments, Inc. 5151 Ward Road Wheat Ridge, Colorado 80033 (303) 422-9112 TELEX: 450681 Head Office: 1 Thorncliffe Park Drive Toronto, Canada M4H1G9 (416) 425-7800 TELEX: 06 23222 EDA TOR Cable: INSTRUMENTS TORONTO

REFERENCES

A. Specifications:

    1. Manufacturer's bulletin, RDA-200 0189.

B. Operations experience:

    1. Moschandreas, D.J., and Rector, H.E., Indoor Radon Concentrations, Environ. Int., vol. 8, pp. 77-82, 1982.

REMARKS

1. Manufacturer states that >1 000 instruments of this type are currently in use in over 40 countries (manufacturer's bulletin).

2. Scintillator efficiency is in excess of 35% or 3.3 cpm/pCi for the RDA-200.

NOTES

RADON
STATIONARY/ACTIVE/ANALYZER

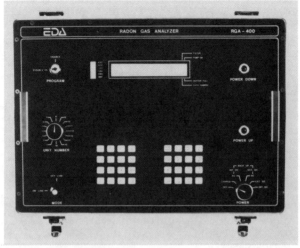

Photograph courtesy of EDA Instruments, Inc.

        Weight: Dimensions:

PRINCIPLE OF OPERATION	Filtration/electrostatic collection/alpha spectroscopy. Ambient air is drawn through a prefilter, which collects radon progeny. Particle-free air enters a coaxially oriented 3 000-mL sample chamber where decay ions are drawn by a strong electrostatic field to be deposited on a solid state detector. Subsequent alpha decay is spectrally analyzed to discriminate between the two Rn isotopes. Radon progeny captured on the prefilter are analyzed in a colinear detection system to discriminate all alpha-emitting progeny. Operation is controlled by a programmable microprocessor; all information is stored in a nonvolatile solid state memory for retrieval.
PERFORMANCE	Lower Detectable Limit: 0.05 pCi/L (gases); 0.002 WL
	Range: 0.0001 to 99.999 WL 0.01 to 99 999 pCi/L 0-99 999 cpm
	Interferences:
	Sampling Rate: 1 L/min, continuous; intervals are selectable
	Accuracy: ±10%
	Reproducibility: ≤5%

R-6
EDA
RGA-400
Radon Gas Monitor
2 of 3

Zero Drift: no inherent drift

Span Drift: no inherent drift

OPERATION

Temperature Range: -10 to 50 °C

Relative Humidity Range: 0 to 100%

Calibration: factory set, no field calibration required

Warmup Time:

Unattended Period: 2 weeks (498 data blocks at selected time intervals)

Maintenance:

Power: 110 V ac, 60 Hz, internal rechargeable standby batteries (up to 10 h backup without data loss); external dc for extended remote applications; 220 V ac, 50 Hz optional

---

FEATURES

Output: 64-character alpha-numeric LCD; optional thermal printer, magnetic cassette recorder, or through CCU 500 (central control unit); output is RS-232 compatible

Training: none required for sampling

Options: DCU-400 thermal printer
DCU-200 magnetic cassette tape recorder
CCU-500 central control unit
RGZ-401 major spare parts kit
WLX-341 filter cartridge (25 per pack)

COSTS

RGA-400: $15 900 (includes console, 25 filter cartridges, ac power cord, external dc power cord, minor spare parts kit, manual)
DCU-400: $ 2 750
DCU-200: $ 3 750
CCU-500: $12 500
RGZ-401: $ 35
WLX-341: $ 65

R-6
EDA
RGA-400
Radon Gas Monitor
3 of 3

MANUFACTURER	EDA Instruments, Inc. 5151 Ward Road Wheat Ridge, Colorado 80033 (303) 422-9112 TELEX: 560681 Head Office: 1 Thorncliffe Park Drive Toronto, Canada M4H1G9 (416) 425-7800 TELEX: 06 23222 EDA TOR Cable: INSTRUMENTS TORONTO
REFERENCES	A. Specifications: 1. Manufacturer's bulletin. B. Operations experience:
REMARKS	1. The unit may be operated as a stand-alone monitor or by using the CCU-500 central control unit. Distributed sampling networks can be formed with central data collection. 2. A high-quality humidity sensor has been incorporated to monitor the relative humidity level of the prefiltered ambient air. The data from the relative humidity sensor are used to compensate for variances in Rn and thoron (Tn) gas concentrations due to detector efficiency changes. The temperature of the prefiltered ambient air is measured as well. 3. The following parameters/functions are available: Rn/Tn gas Rn/Tn ambient PCL Rn/Tn WL INTGRWL Alarm level max Sample interval Date time Spectrum Start data dump Store data dump Data recall

NOTES

Photograph courtesy of EDA Instruments, Inc.

Weight: 6.8 kg   Dimensions: 34 x 30 x 35 cm

PRINCIPLE OF OPERATION	Filtration/alpha detection. Radon progeny are collected on a filter. Alpha activity is detected, averaged, and recorded. Working levels are recorded over periods of 1 h (or 0.1 h, selectable) for up to 41 d. Operation is controlled by an internal microprocessor.

PERFORMANCE

Lower Detectable Limit: 0.0001 WL

Range: 0.0001 to 100 WL

Interferences:

Sampling Rate: 1 L/min, continuous

Accuracy: ±10% (maximum deviation)

Reproducibility: ≤5%

Zero Drift: no inherent drift

Span Drift: no inherent drift

OPERATION

Temperature Range: -10 to 50 °C

Relative Humidity Range: 0 to 100%

Calibration: factory set; no field calibration needed

Warmup Time:

Unattended Period: 41 d

Maintenance: unit is designed for field operation in
              a typically hostile environment; generally
              needs very little maintenance under normal
              operating conditions

Power: 110 ac, 60 Hz, internal rechargeable or 9.5 to 14.0 V
        dc external; 220 ac, 50 Hz optional

FEATURES   Output: five-digit LCD; thermal printer, magnetic cassette
                           recorder, and RS-232 I/O ports

                  Training: none required for sampling

                  Options: DCU-400 thermal printer, ac/dc
                            DCU-040 thermal printer
                            DCU-200 magnetic cassette tape recorder, ac/dc
                            WLX-341 filter discs (25 per pack)
                            WLX-351 15 cm extension legs, set of four
                            WLZ-301 major spare parts kit

COSTS   WLM-300: $6 550 (consists of console, 25 filter discs,
                            ac power cord, minor spare parts kit, manual)
                  DCU-400: $2 750
                  DCU-040: $1 000
                  DCU-200: $3 750
                  WLX-341: $ 60
                  WLX-351: $ 45
                  WLX-301: $ 35

MANUFACTURER   EDA Instruments, Inc.
                  5151 Ward Road
                  Wheat Ridge, Colorado 80033
                  (303) 422-9112
                  TELEX: 450681

                  Head Office:
                  1 Thorncliffe Park Drive
                  Toronto, Canada M4H169
                  TELEX: 06 23222 EDA TOR
                  Cable: INSTRUMENTS TORONTO

REFERENCES

A. Specifications:

   1. Manufacturer's bulletin, WLM-300-0291.

B. Operations experience:

REMARKS

1. Mechanically, the WLM-300 is rugged and environmentally protected to permit operation in hostile outside environments. The unit is lightweight and is powered by line supplies for use in residential or inside industrial applications. For more remote sites, an external 9.5 to 14.0 V dc power may be used or the internal standby batteries may be used to fully operate the unit for up to 10 h.

2. The WLM-300 is simplified by the internal microprocessor, and manual operations amount to replacing the prepackaged filter disc when it becomes loaded, initiating a sampling sequence, and extracting the data from memory. Calibration is unnecessary. The pump is feedback controlled to 1 L/min over the entire operating temperature range for up to 152 cm H_2O back pressures. The operator has full access to the last recorded results through the key pad and a liquid crystal display. Also a number of indicators act as a visual verification of keyed entries and system operation.

NOTES

RADON
STATIONARY/PASSIVE/COLLECTOR

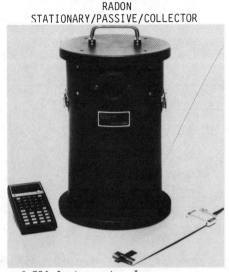

Photograph courtesy of EDA Instruments, Inc.

Weight: 4.5 kg   Dimensions: 35-cm length,
                                                25-cm diameter

PRINCIPLE OF OPERATION	Electrostatic collection/thermoluminescent dosimetry. Ambient Rn diffuses into a chamber where subsequent disintegration of ions are electrostatically focused onto a thermoluminescent dosimeter (TLD) chip held at negative potential in a 900-V electrostatic field. Each alpha particle striking the chip creates metastable defects in the crystal, which can be read and related to integrated Rn concentration. An intervening layer of indicator-quality silica gel and a filter ensure that the air within the sample chamber is desiccated and particle free. A second TLD chip is exposed away from the electrostatic field (in the base of the housing) to check background levels of gamma radiation.

PERFORMANCE   Lower Detectable Limit: 0.03 pCi/L/week (LiF TLD)

                   Range:

                   Interferences:

                   Sampling Rate: diffusion, continuous

                   Accuracy:

                   Reproducibility:

                   Zero Drift:

                   Span Drift:

OPERATION	Temperature Range: <10 to >30 °C
	Relative Humidity Range: 0 to 80% for extended periods
	Calibration:
	Warmup Time: none
	Unattended Period: >1 week (see remark 2)
	Maintenance: replace desiccant as needed, check battery supply voltage
	Power: three Eveready Mini-max No. 493 batteries
FEATURES	Output: nanocoulombs from TLD reader
	Training: none required for sampling
	Options: RDX-727, battery, set of three RDX-351, spare desiccant chamber RDX-458, replacement silica gel 20 cm Whatman #41 filter
COSTS	RDT-310: $575 (complete with batteries, desiccant, special shipping carton, and manual; TLD chips available upon request) RDX-727: $30 each RDX-351: $225 RDX-458: $10/kg Filters: $10
MANUFACTURER	EDA Instruments, Inc. 5151 Ward Road Wheat Ridge, Colorado 80033 (303) 422-9122 TELEX: 450681 Head Office: 1 Thorncliffe Park Drive Toronto, Canada M4H1G9 (416) 425-7800 TELEX: 06 23222 EDA TOR Cable: INSTRUMENTS TORONTO

REFERENCES

A. Specifications:

1. Manufacturer's bulletin.

2. George, A.C., A Passive Environmental Radon Monitor, in Radon Workshop, ed. Breslin, A.J., U.S. Energy Research and Development Administration, Report HASL-325, Health and Safety Laboratory, New York, pp. 25-30, 1977.

B. Operations experience:

REMARKS

1. This instrument is based on the Passive Environmental Radon Monitor (PERM) developed at the DOE Environmental Measurements Laboratory (George 1977).

2. Ordinary unattended exposure periods for this instrument exceed 1 week. The upper limit of exposure is operationally limited by desiccant life, which is determined by humidity. The dessicant can be baked in a home or laboratory oven and reused.

NOTES

RADON
STATIONARY/ACTIVE/ANALYZER

R-9
Harshaw
Radon Daughters
 Analyzer
1 of 2

Photograph courtesy of Harshaw Chemical Company

Weight: 32 kg   Dimensions: detector, 50 x 35 x 40 cm
                                                      computer, 12 x 10 x 10 cm

---

PRINCIPLE OF
OPERATION

Filtration/alpha and beta spectroscopy. Sample air is drawn
through a filter for 2 min. Simultaneously, alpha and beta
backgrounds are measured. The sample deposit on the filter
is transported to the detector where alpha counts (entrance
side of filter) and beta counts (exit side of filter) are
simultaneously registered for 2 min. Radium A and radium C'
are spectroscopically separated by energy. Concentrations
of radium A, radium B, and radium C' plus working levels are
computed automatically.

PERFORMANCE

Lower Detectable Limit: <0.001 WL

Range: <0.001 to 100 WL

Interferences: At extremely high working levels (>100 WL),
                 the resulting gamma background interferes
                 with the performance of the beta detector.

Sampling Rate: 30 to 60 L/min, continuous over 2-min
                 intervals

Accuracy:

Reproducibility: 7% at 10^{-3} WL, 2% at 10^{2} WL

Zero Drift:

Span Drift:

R-9
Harshaw
Radon Daughters
 Analyzer
2 of 2

OPERATION	Temperature Range: -10 to 40 °C
	Relative Humidity Range: no effect
	Calibration: standard source for detectors; flowmeter for sample flow
	Warmup Time: none
	Unattended Period: 1 000 samples (see remark 1)
	Maintenance: exhaust filter should be changed periodically; sample filters can be reused as long as clean
	Power: 110 V ac
FEATURES	Output: eight-digit LCD; thermal printer
	Training: none required for sampling
	Options:
COSTS	$17 000
MANUFACTURER	The Harshaw Chemical Company 1945 E 97th Street Cleveland, Ohio 44106 (216) 721-8300
REFERENCES	A. Specifications:
	1. Manufacturer's bulletin.
	B. Operations experience:
REMARKS	1. With programmed time delays between samples, unattended operation is limited by data storage of 1 000 data points.
	2. Users include EPA, DOE contractors, and a number of State departments of health.

RADON
STATIONARY/PASSIVE/COLLECTOR

Photograph courtesy of Terradex Corporation

Weight: negligible   Dimensions: Type B (total alpha): 6-cm^2 card
                                          Types F, M, and C: 9.5-cm high; 7.3 cm,
                                          widest diameter
                                          Types SF and SM: 2.2-cm high; 3.7 cm,
                                          widest diameter

PRINCIPLE OF   Molecular diffusion/Track Etch™. Alpha particles from Rn in
OPERATION   air or from Rn progeny that have plated out on adjacent
                      surfaces penetrate the detector and cause damage tracks.
                      The damage tracks are chemically etched at the end of the
                      exposure interval and counted. Average exposure is propor-
                      tional to the counted tracks per unit area.

PERFORMANCE   Lower Detectable Limit: 0.16 (pCi/L)-mo (standard). Lower
                                               detectable limits are possible at
                                               an increased cost.

                      Range: 0.16 to 10^4 (pCi/L)-mo

                      Interferences: none

                      Sampling Rate: diffusion, continuous

                      Accuracy: ±1.8 to ±2.8% (relative standard deviation of
                                 calibration factor (Alter and Fleisher 1981))

                      Reproducibility:

                      Zero Drift:

R-10
Terradex
Track Etch™
Radon Detector
2 of 3

	Span Drift:

OPERATION	Temperature Range: -50 to 70 °C
	Relative Humidity Range: 0 to 100%
	Calibration: none required in use
	Warmup Time:
	Unattended Period: depending upon application, <1 mo to >1 yr
	Maintenance: none
	Power: none required for sampling

FEATURES	Output: data report from manufacturer
	Training: none is required for sampling; simple deployment instructions are supplied by manufacturer
	Options: orders may specify "read as needed" to increase sensitivity; see cost section

COSTS	Prices are controlled by number of detectors and desired sensitivity. For types F, M, B, SF, and M detectors, readings at the 4.0 (pCi/L)-mo sensitivity level are $20 each in order lots of 10. The type SW costs $25 each in order lots of 10. All Track Etch™ detectors can be read to increased sensitivity at additional cost.

MANUFACTURER	Terradex Corporation
	460 North Wiget Lane
	Walnut Creek, California 94598
	(415) 938-2545
	TELEX: 337-793

REFERENCES	A. Specifications:
	1. Manufacturer's bulletin.
	2. Alter, H.W., and Fleisher, R.L., Passive Integrating Radon Monitor for Environmental Monitoring, Health Phys., vol. 40, p. 693, 1981.

R-10
Terradex
Track Etch™
Radon Detector
3 of 3

B. Operations experience:

1. Alter, H.W., and Oswald, R.A., Results of Indoor Radon Measurements Using the Track Etch™ Method, Health Phys., vol. 45, no. 2, pp. 425-428, 1983.

2. Prichard, H.M., Gesell, T.F., Hess, C.T., and Weiffenbach, C.V., Associations Between Grab Sample and Integrated Radon Measurements--Dwellings in Maine and Texas, Environ. Int., vol. 8, pp. 83-87, 1982.

REMARKS

1. After the detector has been processed, it is itself a permanent record of the exposure and can be reread at any time. The manufacturer stores the exposed detector for future reference.

2. Shelf life of Track Etch™ detectors is 1 yr if stored in packaging provided by manufacturer.

3. Terradex also produces the type SW detector to measure Rn in water. The device measures the Rn content of air trapped above water in an inverted cup. A known relationship between air concentrations and water concentrations permits a simple calculation of the water concentration.

NOTES

SULFUR DIOXIDE
STATIONARY/ACTIVE/ANALYZER

Photograph courtesy of CEA Instruments, Inc.

Weight: 14 kg   Dimensions: 51 x 41 x 18 cm

PRINCIPLE OF
OPERATION

Automated wet chemistry/colorimetry. Sample air is continuously drawn through distilled water. Absorbed sample
is reacted with pararosaniline and HCHO to form intensely
colored pararosaniline methyl sulfuric acid, whose intensity is measured at 550 nm. Reagent handling and processing is automatic.

PERFORMANCE

Lower Detectable Limit: 0.003 ppm (on 0 to 0.25 ppm full
                                    scale)

Range: 0 to 0.25 ppm (adjustable to 10 ppm)

Interferences: none

Sampling Rate: 250 mL/min, continuous

Accuracy:

Reproducibility: 1%

Zero Drift: <2% for 24 h

Span Drift: <2% for 24 h

OPERATION

Temperature Range: 5 to 48 °C

Relative Humidity Range: 5 to 95%

Calibration: liquid standards, permeation tubes, or
              standard gas dilution

Warmup Time: 20 min

Unattended Period: 18 h on fully charged batteries

Maintenance: tubing in the peristaltic pump should be
              changed once a month

Power: 12 V dc unregulated, 4 W 115/230 V ac, 50/60 Hz

FEATURES

Output: digital panel meter

Training: none required for sampling

Options: stream splitter (to extend range by a factor of
          10 or 100)

COSTS

TGM 555: $5 395
Stream Splitter: $295

MANUFACTURER

CEA Instruments, Inc.
16 Chester Street
P.O. Box 303
Emerson, New Jersey 07630
(201) 967-5660

REFERENCES

A. Specifications:

    1. Manufacturer's bulletin.

B. Operations experience:

REMARKS

SULFUR DIOXIDE
PORTABLE/ACTIVE/ANALYZER

S-2
Interscan
Models 1240
  and 4240
SO_2 Analyzers
1 of 2

Photograph courtesy of Interscan Corporation

Weight: 3.6 kg, Model 1240   Dimensions: 18 x 15 x 29 cm (1240)
        2.0 kg, Model 4240   18 x 10 x 23 cm (4240)

PRINCIPLE OF OPERATION	Electrochemistry. Gas molecules from the moving sample airstream pass through a diffusion medium and are adsorbed onto an electrocatalytic sensing electrode where subsequent reactions generate an electric current. The diffusion-limited current is linearly proportional to SO_2 concentration.

PERFORMANCE

Lower Detectable Limit: 1% of full scale (0.010 ppm)

Range: 0 to 1 ppm, 0 to 5 ppm, 0 to 10 ppm (other ranges available)

Interferences: expressed as ppm of interferent needed to give 1 ppm deflection: NH_3 = 45, NO_2 = 10. C_2H_5SH, H_2S, and CH_3SH require a special scrubber element.

Sampling Rate: continuous

Accuracy: ±2% of full scale

Reproducibility: ±0.5%

Zero Drift: ±1% full scale in 24 h

Span Drift: <±2% full scale in 24 h

S-2
Interscan
Models 1240
 and 4240
SO$_2$ Analyzers
2 of 2

OPERATION	Temperature Range: 10 to 120 °F
	Relative Humidity Range:
	Calibration: standard gas mixture
	Warmup Time:
	Unattended Period: 10 h on battery power
	Maintenance:
	Power: four alkaline MnO$_2$ batteries for amplifier, two Ni-Cd for pumps LCD; one HgO battery for bias amplifier reference
FEATURES	Output: 0 to 100 mV full scale
	Training: none required for sampling
	Options: 1240, audible and visual alarm 4240, audible alarm
COSTS	Model 1240: $1 675 Model 4240: $1 895
MANUFACTURER	Interscan Corporation P.O. Box 2496 21700 Nordhoff Street Chatsworth, California 91311 (213) 882-2331 TELEX: 67-4897
REFERENCES	A. Specifications: 1. Manufacturer's bulletin. B. Operations experience:
REMARKS	1. These units were originally designed for workplace monitoring (TLV = 2 ppm) and may be of limited use in nonindustrial environments unless high concentrations are present.

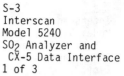

SULFUR DIOXIDE
PERSONAL/PASSIVE/ANALYZER

S-3
Interscan
Model 5240
SO$_2$ Analyzer and
  CX-5 Data Interface
1 of 3

Photograph courtesy of Interscan Corporation (Model 5140 shown is identical in appearance to Model 5240)

Weight: 0.7 kg (5240)   Dimensions: 15 x 8 x 5 cm (5240)
          0.6 kg (CX-5)   7 x 18 x 13 cm (CX-5)

PRINCIPLE OF   Diffusion/electrochemistry. SO$_2$ diffuses into an electro-
OPERATION   chemical cell, producing a signal proportional to SO$_2$
                      concentrations. The signal is digitized, incorporated into
                      1-min averages, and stored. Nondestructive recovery of each
                      1-min average is accomplished through a separate data reader.
                      Data storage capacity is 2 048 1-min averages. Stored data
                      can be retrieved through the CX-5 interface and transferred
                      to a computer for further processing.

PERFORMANCE   Lower Detectable Limit: 0.5% of full scale (0.010 ppm)

                     Range: 0 to 20 ppm

                     Interferences: expressed as ppm of interferent needed to
                                      give 1 ppm deflection: NH$_3$ = 45, NO$_2$ = 10.
                                      Special filters are required for ethyl
                                      mercaptans, methyl mercaptans, and H$_2$S when
                                      these interferents are at concentrations
                                      equivalent to SO$_2$.

                     Sampling Rate: diffusion, continuous

                     Accuracy: ±2% of reading, ±1 least significant digit (LSD),
                                ±0.5% of full-scale reading

                     Reproducibility: ±1% reading, ±1 LSD

S-3
Interscan
Model 5240
SO$_2$ Analyzer and
  CX-5 Data Interface
2 of 3

	Zero Drift: ±1% reading, ±1 LSD in 24 h
	Span Drift: ±1% reading, ±1 LSD in 24 h
OPERATION	Temperature Range: 30 to 120 °F
	Relative Humidity Range: 1 to 100%
	Calibration: standard gas mixture
	Warmup Time: <5 min
	Unattended Period: up to 34 h
	Maintenance: calibration, battery replacement, sensor replacement
	Power: long life 9-V battery (alkaline MnO$_2$, NEDA type 1604A); battery life is 125 h continuous operation
FEATURES	Output: printout from data reader (see remark 2)
	Training: none required for sampling
	Options:
COSTS	Model 5240: $1 145 CX-5: $1 375
MANUFACTURER	Interscan Corporation P.O. Box 2496 21700 Nordhoff Street Chatsworth, California 91311 (213) 882-2331 TELEX: 67-4897
REFERENCES	A. Specifications:
	1. Manufacturer's bulletin.
	B. Operations experience:

S-3
Interscan
Model 5240
SO_2 Analyzer and
  CX-5 Data Interface
3 of 3

REMARKS

1. The CX-5 interface allows nondestructive retrieval of all 1-min averages stored in the dosimeter memory. The interface is programmed to give time-selected and time-weighted averages, maximum short-term exposure levels, and the time at which they occur. Peak concentration and the time at which it occurs is also given. One CX-5 can service many 5000 series dosimeters.

2. Data readout may also be accomplished by a device available from:

   Metrosonics, Inc.
   P.O. Box 23075
   Rochester, New York 14692
   (716) 334-7300

3. These units were originally designed for workplace monitoring (TLV = 2 ppm) and may be of limited use in nonindustrial environments unless high concentrations are present.

NOTES

EPA REFERENCE AND EQUIVALENT METHODS

Continuous analyzers for CO, NO_2, SO_2, and O_3 that appear on EPA's "List of Designated Reference and Equivalent Methods" are enumerated here. Table A-1 displays the performance specifications for SO_2, O_3, CO, and NO_2. Specific instruments that have been designated reference or equivalent for each pollutant are listed in Table A-2. Addresses and telephone numbers of manufacturers are contained in Table A-3.

TABLE A-1. Performance Specifications for Automated Methods*

	Performance parameter	Units	Sulfur dioxide	Ozone	Carbon monoxide	Nitrogen dioxide
1.	Range	ppm	0-0.5	0-0.5	0-50	0-0.5
2.	Noise	ppm	0.005	0.005	0.50	0.005
3.	Lower detectable limit	ppm	0.01	0.01	1.0	0.01
4.	Interference equivalent					
	Each interferent	ppm	±0.02	±0.02	±1.0	±0.02
	Total interferent	ppm	±0.06	±0.06	±1.5	±0.04
5.	Zero drift, 12 and 24 h	ppm	±0.02	±0.02	±1.0	±0.02
6.	Span drift, 24 h					
	20% of upper range limit	%	±20.0	±20.0	±10.0	±20.0
	80% of upper range limit	%	±5.0	±5.0	±2.5	±5.0
7.	Lag time	min	20	20	10	20
8.	Rise time	min	15	15	5	15
9.	Fall time	min	15	15	5	15
10.	Precision					
	20% of upper range limit	ppm	0.01	0.01	0.5	0.02
	80% of upper range limit	ppm	0.015	0.01	0.5	0.03

*In accordance with 40 CFR part 53, Federal Register, vol. 40, p. 7049, February 18, 1975 (as amended vol. 40, p. 18168, April 25, 1975; vol. 41, p. 52694, December 1, 1976) see also Purdue, L.J., EPA Reference and Equivalent Methods, J. Air Pollut. Control Assoc., vol. 30, no. 9, pp. 992-996, 1980.

TABLE A-2. Summary of Commercially Available Instruments for U.S. Environmental Protection Agency Designated Reference and Equivalent Methods for CO, NO_2, SO_2, and O_3 (Parenthetical values indicate approved ranges.)

Pollutant	Methods
Carbon monoxide	**Reference methods** Nondispersive infrared (NDIR) - Bendix 8501-5CA (50) - Beckman 866 (50) - Horiba AQM-10-11 and 12 (50) - Horiba 300E/300SE (20, 50, 100) - Monitor Labs 8310 (50) - MSA 202S (50) Gas filter correlation (GFC) - Dasibi 3003 (50) - Thermo Electron 48 (50)
Nitrogen dioxide	**Reference methods** Gas-phase chemiluminescence - Beckman 952A (0.5) - Bendix 8101-B and C (0.5) - CSI 1600 (0.5) - Meloy NA530R (0.1, 0.25, 0.5, 1.0) - Monitor Labs 8440E (0.5) - Monitor Labs 8840 (0.5, 1.0) - Phillips PW 9762/02 (0.5) - Thermo Electron 14 B/E and D/E (0.5)
Ozone	**Reference methods** Gas phase chemiluminescence - Beckman 950A (0.5) - Bendix 6002 (0.5) - CSI 2000 (0.5) - McMillan 1100-1, 2, and 3 (0.5) - Meloy OA325-2R and OA350-2R (0.5) - Monitor Labs 8410E (0.5) **Equivalent methods** Ultraviolet absorption - Dasibi 1003-AH, -PC, -RS (0.5, 1.0) - Monitor Labs 8810 (0.5, 1.0) - PCI Ozone Corp LC-12 (0.5) - Thermo Electron 49 (0.5, 1.0) Gas-solid phase chemiluminescence - Phillips PW 9771 (0.5)

(Continued)

TABLE A-2. Summary of Commercially Available Instruments for U.S. Environmental Protection Agency Designated Reference and Equivalent Methods for CO, NO$_2$, SO$_2$, and O$_3$ (Parenthetical values indicate approved ranges.) (Concluded)

Pollutant	Methods
Sulfur dioxide	**Equivalent methods** Flame photometric detection (FPD) - Bendix 8303 (0.5, 1.0) - Meloy SA185-2A (0.5, 1.0) - Meloy SA285E (0.05, 0.1, 0.5, 1.0) - Monitor Labs 8450 (0.5, 1.0) Pulsed ultraviolet fluorescence - Beckman 953 (0.5, 1.0) - Lear Siegler AM2020 (0.5) - Meloy SA700 (0.25, 0.5, 1.0) - Monitor Labs 8850 (0.5, 1.0) - Thermo Electron 43 (0.5, 1.0) Second derivative spectroscopy - Lear Siegler SM1000 (0.5) Automated wet chemical - Phillips PW9755 and PW9700 (0.5)

TABLE A-3. Manufacturers of Stationary Analyzers that Appear in Table A-2

Beckman Instruments
Process Instruments Division
2500 Harbor Boulevard
Fullerton, California 92634
(714) 871-4848

Bendix Corp.
Environmental and Process
 Instruments Division
Box 831
Lewisburg, West Virginia 24901
(304) 647-4358

Columbia Scientific Industries
Box 9908
Austin, Texas 78766
(512) 258-5191
(800) 531-5003

Dasibi Environmental Corp.
616 East Colorado Street
Glendale, California 91205
(213) 247-7601

Horiba Instruments, Inc.
1021 Duryea Avenue
Irvine, California 92714
(714) 540-7874

Lear Siegler, Inc.
74 Inverness Drive E
Englewood, Colorado 80112
(303) 770-3300

MSA
600 Penn Center Boulevard
Pittsburgh, Pennsylvania 15235
(412) 273-5172

Monitor Labs, Inc.
10180 Scripps Range Boulevard
San Diego, California 92131
(619) 578-5060

PCI Ozone Corp.
One Fairfield Crescent
West Caldwell, New Jersey 07006
(201) 575-7052

Phillips Electronic Instruments
85 McKee Drive
Mahwah, New Jersey 07430
(201) 529-3800

Thermo Electron Corp.
Environmental Instruments Division
108 South Street
Hopkinton, Massachusetts 01748
(617) 435-5421

Appendix B

ALTERNATIVES TO COMMERCIAL INSTRUMENTATION: USER-CONFIGURED METHODS

INTRODUCTION

User-configured sampling systems sometimes represent a viable alternative to commercially available instruments. Such systems can be cost effective without sacrificing data quality. This appendix summarizes user-configured methods for (1) asbestos and other fibrous aerosols, (2) formaldehyde (HCHO), (3) inhalable particulate matter (IP), (4) nitrogen dioxide (NO_2), (5) organic pollutants, and (6) radon (Rn). Methods described here represent either practices endorsed by an appropriate organization or widely accepted techniques found in refereed professional journals. Decisions to implement any of these methods should be made only after a careful review of the supporting literature. Even then, a full range of pilot tests should be carried out prior to formal sampling to verify performance.

ASBESTOS AND OTHER FIBROUS AEROSOLS

Airborne fibers of interest in indoor air quality monitoring include all varieties of asbestos; a number of manufactured fibers such as mineral wools, fibrous glass wool, and some ceramics; and organic fibers such as animal dander and wood dusts (U.S. Environmental Protection Agency (EPA) 1981). Manual sampling for fibrous aerosols generally consists of drawing sample air through a membrane filter. The number of fibers in the collected sample can be counted using optical microscopy. Greater specificity in terms of type of fibers can be obtained by using more sophisticated procedures such as electron microscopy and X-ray diffraction.

Optical Microscopy

Detailed procedures for determining exposure to airborne asbestos fibers though filter collection and optical sizing and counting are available in National Institute for Occupational Safety and Health (NIOSH) P&CAM 239 (NIOSH 1977). In this method, asbestos fibers are defined as particles of physical dimension greater than 5 μm with a length-to-diameter ratio of 3:1 or greater. This method is not asbestos specific; rather, it assesses all fibers that meet these size criteria. Furthermore, the resolution limits of optical microscopy and the assigned cutoff of 5 μm preclude assessing fibers that fall below this size range. Thus, the method provides an index of asbestos exposure rather than a true measure of asbestos fiber counts.

Sampling is usually carried out using a 34-mm membrane filter with 0.8 μm pore size mounted in an open-faced cassette. Sample flow is selected upon considering the desired sampling period and minimum detection limits. For personal monitoring applications, a number of battery-powered pumps with stable flow-rate control are available.

At the end of sampling, exposed filters are resealed in their cassettes and taken to the laboratory where they are sectioned, mounted onto microscope slides, and, through chemical treatment, made transparent. The slides are then examined using phase-contrast illumination at a magnification of 400 X to 500 X to count fibers.

Though NIOSH indicates that the experience level of the analyst performing the fiber count does not significantly contribute to variations, individuals who work only from published instructions often obtain fiber counts that are as little as half of those obtained by trained and experienced fiber counters (Sawyer and Spooner 1978). Therefore, formal training in optical microscopy is necessary. Introductory and continuing training programs are offered by a number of organizations; published notices frequently appear in periodicals such as the Journal of the Air Pollution Control Association and the American Industrial Hygiene Association Journal.

A number of commercial laboratories offer analytical services for user-supplied filter samples. The American Industrial Hygiene Association maintains a laboratory accreditation program and periodically publishes a list of accredited laboratories in its journal. Accredited laboratories are required to participate in the NIOSH Proficiency Analytical Testing (PAT) program.

Electron Microscopy and X-Ray Diffraction

Greater specificity and accuracy in fiber measurements can be gained by examining the morphology and structure of individual fibers through electron microscopy or by determining fiber composition through X-ray diffraction. As with optically analyzed samples, fibers are collected onto membrane filters; however, sample preparation is much more complex, and the analytical equipment requires trained personnel (Sawyer and Spooner 1978). These techniques, though much more expensive than optical counting, do address fibers of all sizes and provide a mineral-specific exposure assessment.

BIOLOGICAL AEROSOLS

Sampling methods for biological aerosols are essentially no different from those employed for particulate matter. As summarized in Chatigny (1983), Chatigny et al. (1983), and Solomon (1984), four categories of sampling principles are in use: (1) gravitation settling, (2) filtration, (3) liquid impingement, and (4) inertial impaction.

Fallout samplers, which may be simply open petri dishes or glass slides, provide rather limited information because the sample deposit is not easily related to airborne concentrations and collection efficiency is heavily skewed toward larger particles. Filtration samplers and liquid impingement samplers are volumetric samplers, as are inertial impactors. Inertial impactors provide for collection based on aerodynamic size (see instrument summaries B-1 and B-2 in Appendix A).

The immense variety of viable and nonviable particles, as well as the lack of automated analytical techniques, is what sets biological aerosols apart from the other indoor air pollutants discussed in this book. For viable aerosols, culture methods are widely used to identify microorganisms. Optimum culture media and growing conditions vary from organism to organism, requiring multiple samples when a variety of microorganisms is being investigated. Furthermore, there are no practical identification methods for a number of fungal spores, algae, actino- mycetes and bacteria; the same holds true for certain types of biogenic debris.

The selection of sampling and analytical methods for biological aerosols has generally reflected investigator preferences tempered by available equipment. The American Conference of Governmental Industrial Hygienists (see Chapter 7) has formed a committee to develop standardized protocols for biological aerosols.

FORMALDEHYDE

The most popular methods for measuring indoor HCHO concentrations employ impingers for collection in an absorbing reagent for air sampling followed by colorimetric analysis in a laboratory. For better sampling efficiencies, two bubblers in series, operating under vacuum, are recommended. The collection efficiency of one bubbler is approximately 80 percent; the second bubbler boosts the total collection efficiency to approximately 95 percent. The contents of each bubbler may be analyzed separately or the contents may be pooled. Additionally, to reduce the loss, when the absorbing reagent is water, sampling frequently takes place with the bubblers chilled. A vapor trap, which is simply an empty bubbler, should be installed after the second bubbler to protect the pump.

Chromotropic Acid Method

The modified NIOSH impinger/chromotropic method is the most widely used sampling and analysis method for HCHO measurements (Godish 1985). It differs from NIOSH P&CAM 125 (NIOSH 1977) in that the sample air is bubbled through a 1-percent sodium bisulfite solution in place of distilled water. In the laboratory, chromotropic acid reagent is added to an aliquot of the absorbing solution. Concentrated sulfuric acid is added slowly to the absorbing solution to avoid spattering due to the exothermic reaction. The treated aliquot is allowed to cool to room temperature. Absorbance is read at 580 nm in a spectrophotometer. HCHO content is determined from a curve derived from fresh standard HCHO solutions.

Concentrations as low as 0.1 ppm can be determined in a 25-L air sample. (This lower detection limit is based on 20 mL of absorbing solution and a difference of 0.05 absorbance units above blank.) Sensitivity can be enhanced by increasing the sample air volume. This increase can be accomplished by extending the sample period, increasing the flow rate, or adjusting the amount of absorbing solution in the bubblers. Godish (1981, 1985) recommends a sample flow rate of 1 L/min, a minimum 90-min sample period, and 10-mL absorbing reagent in each impinger.

Modified Pararosaniline Method

Sample air is bubbled through deionized, distilled water that is kept chilled by an ice bath or refrigeration during sampling. In the laboratory, acidified pararosaniline is added to an aliquot of the sample solution and thoroughly mixed. Then sodium sulfite reagent is added and the solution is again thoroughly mixed. The treated aliquot is placed in a 25 °C water bath for 60 min to allow color development. Absorbance is read at 570 nm in a spectrophotometer. HCHO content is determined from a curve derived from fresh standard solutions. Detailed proce- dures for the modified pararosaniline method may be found in Miksch et al. (1981).

Experiments conducted by Daggett and Stock (1985) showed rapid degradation of stored HCHO in water samples. Under ambient temperatures, up to 80 percent of the HCHO was lost over a 72-h period. Under refrigerated storage, losses were smaller (10 to 15 percent). They recommend that samples be analyzed within 24 h or frozen immediately if longer storage is required.

HCHO concentrations as low as 0.025 ppm can be determined in a 60-L air sample. (This lower detection limit is based on 20 mL of absorbing solution and a difference of 0.05 absorbance units above blank.) Thus, this method is twice as sensitive as the modified chromotropic method but it requires refrigeration of impingers while sampling.

Other Methods

A visual colorimetric screening method based on the methyl-2-benzothiazolinone hydrazone (MBTH) technique has been reported by Matthews and Howell (1981). This method is specific for all aliphatic aldehydes, but in domestic indoor air settings, HCHO is expected to be the principal contributor. Sampling is carried out using a passive semipermeable membrane collector with water as an absorbent. When the color change is fully developed, the solution is compared to a reference color chart to determine concentration range.

HCHO concentrations may also be determined by collection onto various solid sorbents followed by laboratory analysis. Beasley et al. (1980) suggest collection onto silica gel coated with 2,4-dinitrophenylhydrazine (2,4-DNPH). During sampling, HCHO forms a specific hydrazone that is extracted using acetonitrile and quantified by high performance liquid chromatography (HPLC) with ultraviolet detection.

Matthews and Howell (1982) have developed a simple approach using 13 X molecular sieve collection followed by water-rinse desorption and colorimetric analysis based on the modified pararosaniline method. This approach has been tested for passive sampling as well as active sampling with a pump and has shown high collection efficiencies (>99.9 percent) and stability. The shelf life of sealed exposed media at <38 °C is at least 1 week; less than 10 percent degradation was observed after a month. The desorption efficiency for HCHO is 92 percent. The lower detection limit (based on a 30-L air sample using 5 g of sorbent) is 0.025 ppm HCHO. Care must be exercised in applying this technique, however, because the sorbent also has an affinity for water; one study suggested limiting sampling to 2 L/min for 30 min (Battelle 1982).

INHALABLE PARTICULATE MATTER

For indoor and personal monitoring, the ability to measure not only the mass of IP but also the size distribution and chemical or elemental composition is crucial. Sulfates, nitrates, and a number of metals such as lead (Pb) need to be reliably determined using quiet, rugged personal monitors. However, no commercial instruments are available that meet all of these requirements.

A number of user-configurable methods are available to collect indoor samples of IP and other size fractions (Fletcher 1984). There are very few publications, however, that provide sufficient information to fabricate in-house alternatives to commercial instruments. One such system, designed and tested by the Harvard University School of Public Health under Electric Power Research Institute (EPRI) sponsorship (Turner et al. 1979a,b), uses a miniature cyclone to separate the respirable fraction (<3.5 μm) for filter collection.

Another system developed by the National Bureau of Standards under EPA (NBS/EPA) sponsorship (Bright and Fletcher 1983), collects two size fractions--fine (<3 μm) and coarse ($\geq$3 μm); separate inserts for the sampling head allow an upper-size limit of 7, 10, or 15 μm. Though not commercially available at this time, both systems can be assembled from readily available components and materials. The NBS/EPA sampling system requires some machining. The inlets are aluminum and machining needs can be met on a standard lathe. Specifications for constructing the NBS/EPA inlet are described by Bright and Fletcher (1983).

Early in the development process, both research groups recognized a need to improve upon commercially available pumps to ensure acceptable performance. Both of the articles cited above provide a useful chronicle of this important feature.

The Harvard University School of Public Health has also developed a single-stage aerosol impactor especially suited for indoor air quality studies (Turner et al. 1984). The system can be configured to measure either IP (<10 μm) or the fine fraction (<2.5 μm) at 4 L/min. In preliminary testing performed by GEOMET (Nagda et al. 1985), the NBS/EPA sampling system and the Harvard single-stage impactor were found to give highly reproducible results under a variety of conditions.

Inorganic Analysis of Particulate Matter Samples

The inorganic constituents of particulate matter that are of interest to indoor air quality studies include sulfates, nitrates, and a number of metals. Among the metals, principal interest has focused on Pb, though all metallic and semi-metallic elements may be of interest. Inorganic material is usually collected in conjunction with gravimetric sampling for IP. Constituents of interest may be extracted for quantitation, or the sample matrix may be submitted directly to analysis.

For any given constituent, a wide variety of methods and tested procedures are available; extensive summaries can be found in Katz (1977, 1980). These methods range from relatively simple approaches involving extraction and spectrophotometric determination for compounds such as sulfates and nitrates to the more elaborate approaches of atomic absorption spectroscopy, neutron activation analysis, proton-induced X-ray emission (PIXE), and X-ray fluorescence for metals.

In many cases, methods for inorganic constituents were developed initially for source testing and for high-volume sampling. With the smaller sample masses that are generally captured with size-selective samplers, and particularly with the low flow rates used for personal samplers, attention should be given to the manner in which analytical performance interacts with sample mass; that is, detection limits of the analytical method must correspond to lowest concentrations of interest.

NITROGEN DIOXIDE

Passive samplers for NO_2 can be readily fabricated through the method reported by Palmes et al. (1976). Constructed entirely from commercially available materials, the Palmes tube provides for integrated exposure measurement with a compact and lightweight sampler.

Sampling rate is controlled by molecular diffusion. Wafers of stainless steel screening coated with triethanolamine (TEA), a highly efficient NO_2 collector, are positioned at the closed end of a hollow tube. The TEA maintains a near-zero NO_2

concentration at the closed end of the tube. The concentration gradient between the environment and the sorbent causes a net flow of NO_2 to the TEA in accordance with Fick's First Law of Diffusion.

NO_2 in the environment surrounding the tube enters the tube while the open end of the tube is uncapped and is trapped by the TEA. Recapping the tube interrupts the concentration gradient and halts sampling. The cap for the free end of the Palmes tube should be clearly distinguishable from the cap retaining the TEA-coated screens.

The amount of NO_2 collected on the TEA depends on the concentration as well as the duration of the sampling period. Thus, with this sampler, an exposure to 1 ppm NO_2 for 1 h gives the same sample mass as an exposure to 0.5 ppm for 2 h.

Sample analysis is spectrophotometric, involving a reagent system of N-1-naphthylethylene-diamine-dihydrochloride (NEDA) and sulfanilamide that reacts with the trapped NO_2 to form a purplish chromophore. The absorbance, read at 540 nm, follows Beer's law, giving approximately 1 absorbance unit per 40 nmol of NO_2 in 2.1 mL of reagent. Standard curves using sodium nitrite solutions need to be prepared for each analytical session.

The method generally is not subject to chemical interferences. According to diffusion theory, there is a slight (1 or 2 percent per 10 °C) change in sample collection rate due to temperature changes. Girman et al. (1983) have noted temperature-related changes approaching 15 percent (1 percent per °C over the range 15 to 27 °C) that appear to be related to a change of phase for TEA at 21 °C.

The working range of 0 to 40 nmol allows up to 17 ppm h exposure (approximately 100 ppb average NO_2 concentration for a week). Higher exposures can be accom-modated in the laboratory through dilution. The Palmes tube was originally developed for workplace monitoring where levels of concern are in the parts per million range and sampling periods are a few hours. For indoor air quality moni-toring, levels of concern are much lower; the most common sample period is 1 week.

ORGANIC POLLUTANTS

Relatively limited work has been carried out to characterize organic pollutants in the indoor environment. It is a highly complex topic and the list of airborne organic compounds of interest is large. Their presence has been attributed to the use of solvents or solvent-containing products, to emanations from manu-factured materials, and to combustion products.

The majority of the monitoring approaches involves selectively concentrating tar-get compounds on a collector, such as a sorbent bed or filter, and transferring the sample to the laboratory for analysis. Continuing advances in analytical methods such as chromatography and mass spectrometry permit reliable detection and speciation from submicrogram quantities on a routine basis.

Comprehensive reviews of methods that may be adapted to indoor settings are to be found in Lamb et al. (1980), in Katz (1980), in Riggin (1983), and in Fox (1985). Organic pollutants may be classified into three broad classes:

● Volatile organic compounds (VOC)--relatively low molecular weight species that exist in the vapor phase under ordinary ambient conditions. Examples include benzene and carbon tetrachloride.

- Semivolatile organic compounds (SVOC)--less volatile species such as polychlorinated biphenyls and pesticides.

- Organic aerosols--higher weight molecular species that usually exist in the liquid or solid phase under ordinary conditions. Examples include a wide range of polynuclear aromatic hydrocarbons condensed onto particulate matter.

Volatile Organic Compounds

VOCs are ordinarily collected by drawing sample air through a sorbent bed that traps and retains target compounds. The VOCs of interest are later desorbed in the laboratory and quantitated. Sampling trains can be configured to meet the needs of monitoring strategies requiring personal monitoring, indoor fixed monitoring, or outdoor fixed monitoring.

A number of solid sorbents are available. Tenax-GC and activated charcoal have been widely used to sample VOCs. Tenax-GC is probably the most widely used solid sorbent; it is not, however, regarded as adequate for highly volatile substances (Krost et al. 1982). Some researchers suggest a combination of stainless steel canisters and solid sorbents to capture the range of volatilities (Cox 1983). The particular advantage that Tenax and other porous polymers offer is the thermal stability to allow desorption at high temperatures (up to 300 °C). Charcoal has been questionable in this regard.

Captured material may be desorbed using solvent elution, and an aliquot may be injected into a gas chromatograph (GC) for quantitation. Solvent elution, however, partially offsets the advantages of sorbent trapping by rediluting the sorbent-concentrated sample. Solvent purity is also critical; trace level contaminants in the solvent can corrupt the sample unless extreme care is exercised.

Breakthrough and inherent limits of detection of the analytical system are of prime importance in considering the use of solid sorbents in sampling VOCs. Breakthrough refers to saturation capacity of the sorbent bed so that elution occurs during sampling, and subsequent quantitation could severely underestimate concentrations. Breakthrough volume (that is, the volume of air sampled beyond which more than 50 percent of a particular target compound entering the front of the sampling cartridge is lost in the exit stream of sampled air) is a useful concept in determining optimum sample volumes and the size of a sorbent bed needed to meet the detection limit of the analytical system. Breakthrough volumes can vary considerably from compound to compound; determinations of sampling rates should rely on literature estimates as well as experience.

Collector tubes can be configured from glass or stainless steel. The inlet and outlet are usually plugged with glass wool to provide support. Extreme care must be exercised in all handling of the sorbent to preclude contamination. Usual precautions include extended Soxhlet extraction of virgin and reused Tenax, followed by vacuum drying at 100 °C. Mesh sorting and tube packing are performed under clean room conditions. Storage and handling when not actively sampling should be through the use of clean, sealed containers. A number of manufacturers such as SKC and Perkin-Elmer market preassembled sorbent tubes that can be analyzed by a commercial laboratory. Vacuum sources are available to allow configurations for personal and fixed sampling over desired time periods (Wallace and Ott 1982).

Portable chromatographs are available to support near real-time data collection in the field. Such systems can be configured to analyze large numbers of locally collected samples to guide further sampling, or, with accessories such as automatic sample loops and data logging, such systems can be configured to automatically sample environments on a time base of a few minutes. Discussions of basic strategies and equipment may be found in Clay and Spittler (1982), Spittler et al. (1983), and Riggin (1983).

Semivolatile Organic Compounds

SVOCs--particularly polychlorinated biphenyls and pesticides--are generally sampled by drawing sample air through polyurethane foam (called PUF or PFOAM). Target compounds are subsequently extracted in the laboratory and analyzed.

Lewis and MacLeod (1982) have developed an approach that is adaptable to personal monitoring as well as fixed sampling. Cylindrical PUF plugs (22-mm diameter x 76 mm) are cut from sheet stock, open-cell, polyether-type polyurethane foam (0.022 g/cm^3) that is ordinarily used for upholstery. Cylindrical cuts are facilitated by a stainless steel cutting die. Preparation requires extended Soxhlet extraction using hexane to remove contaminants from the PUF plugs. Plugs are inserted into a borosilicate glass tube (20-mm inner diameter x 80 mm), one end of which is drawn down to a 7-mm outer diameter, open connection to allow attachment of a vacuum line.

Sample cartridges should be protected from contamination by wrapping in hexane-washed aluminum foil. Nominal sample volumes may be as high as 3 m^3 (that is, 4 L/min for 12 h, for instance). PUF plugs are removed from the cartridges in the laboratory, target compounds are extracted with a Soxhlet extractor using diethylether in hexane, and quantitation may be carried out using GC or HPLC. Compounds of interest can be measured at levels as low as 1 ng/m^3.

Sampling can be extended to include VOCs and at the same time fortify efficiency for some SVOCs by inserting a Tenax sandwich between two shortened PUF plugs in a single cartridge.

Organic Aerosols

In many cases, sampling for organic aerosols is done in conjunction with standard gravimetric sampling (i.e., as for IP). Attention in this area has focused largely on polynuclear aromatic hydrocarbons (PAHs), which represent ubiquitous combustion products that have demonstrated animal carcinogenicity.

Once extracted from the filter by an appropriate solvent, PAHs may be quantified as a class or as individual compounds by a variety of methods including routine GC, HPLC, and thin-layer chromatography (TLC) (Katz 1980).

Spectroscopic techniques have also come into use. In particular, room-temperature phosphorescence (RTP) spectroscopy has overcome the need for cryogenic treatment (Vo-Dinh et al. 1981). In this method, PAHs are isolated by liquid chromatography and diluted in ethanol. A 3 µL aliquot is delivered onto a filter previously treated with a solution of heavy metal salts and irradiated with ultraviolet light. Resulting phosphorescence is enhanced by the heavy metals and can be compared to reference spectra levels to identify and quantify PAHs. Continuing work in RTP spectroscopy has resulted in a passive monitor for PAH vapors (Vo-Dinh 1985). The dosimeter can detect a variety of PAH compounds at ppb levels from 1-h samples.

In many cases analytical methods for PAHs were initially developed for source testing and for high-volume ambient sampling. With the smaller sample masses that are generally captured with size selection and particularly with personal monitors for particulate matter, attention should be given to the manner in which analytical performance interacts with sample mass. For example, detection of nanogram quantities may well force collection volumes beyond the flow capacity of the sampler. Or worse still, a short sampling period coupled to a large sample volume could severely alter airflow patterns in some indoor settings. In many cases, some compromise can be achieved by reducing the extraction volume or by pooling replicate samples for composite analysis.

RADON

Using commercially available passive devices, average indoor Rn concentrations can be measured over periods of several months using the Track Etch™ Method (Alter and Fleischer 1981) or over periods of a few weeks using thermoluminescence dosimetry (George and Breslin 1977). For shorter periods (i.e., on the order of a few days), no commercially available passive devices have been identified. George (1984), however, has reported a passive method based on activated carbon and gamma ray detection that is relatively inexpensive, maintenance free, and easily fabricated from commercially available components. The activated carbon canister method exhibits a lower limit of detection for Rn of 0.2 nCi/m^3 for an exposure period of 72 h.

The device, originally based on the M11 canister developed by the U.S. Army Chemical Corps in World War II, is a cylindrical container 2.5-cm high by 10-cm in diameter, which is filled with approximately 100 g of activated carbon to a depth of 2 cm. A metal screen and retainer ring hold the activated carbon in place. The canister is fitted with a removable metal cover taped in place to provide an airtight seal when not sampling.

To sample, the metal cover is simply removed in the area to be monitored and resealed at the end of the sampling period. The amount of sorbed Rn in the carbon bed is determined by measuring the total gamma activity produced by the decay of Pb-214 (0.242, 0.294, and 0.352 MeV) and Bi-214 (0.609 MeV). The gamma activity of exposed canisters can be measured up to 10 days after the end of exposure, thus allowing the devices to be transferred by mail.

Calibration tests showed no discernible differences in collection efficiency for temperatures ranging from 18 to 27 °C. Response of the device, however, decreases with exposure time and humidity. Corrections for humidity can be made by gravimetrically determining the amount of water sorbed during exposure and applying a second calibration curve.

These devices are reusable. Before reuse, residual Rn and sorbed water can be purged with heated air (100 °C for 5 min) or by baking in an oven at 120 °C for several hours.

REFERENCES

Alter, H.W., and Fleischer, R.L., Passive Integrating Radon Monitor for Environmental Monitoring, Health Phys., vol. 40, p. 693, 1981.

Battelle, Quality Assurance Project Plan for Control Technology Assessment and Exposure Profile for Workers Exposed to Hazards in the Electronic Components Industry, U.S. Environmental Protection Agency, Contract Number 68-03-3026, Battelle Columbus Laboratories, Columbus, 1982.

Beasley, R.K., Hoffman, C.E., Rueppel, M.L., and Warley, J.W., Sampling Formaldehyde in Air With Coated Solid Sorbent and Determination by High Performance Liquid Chromatography, Anal. Chem., vol. 52, ed. 7, pp. 1110-1114, 1980.

Bright, D.S., and Fletcher, R.A., New Portable Ambient Aerosol Sampler, Am. Ind. Hyg. Assoc. J., vol. 44, no. 7, pp. 528-536, 1983.

Chatigny, M.A., Sampling Airborne Microorganisms, Air Sampling Instruments, 6th ed., eds. P.J. Lioy and M.J.Y. Lioy, American Conference of Governmental Industrial Hygienists, Cincinnati, pp. E-1 to E-9, 1983.

Chatigny, M.A., Wolochow, H., and Hinton, D.O., Sampling Aeroallergens, Air Sampling Instruments, 6th ed., eds. P.J. Lioy and M.J.Y. Lioy, American Conference of Governmental Industrial Hygienists, Cincinnati, pp. F-1 to F-5, 1983.

Clay, P.F., and Spittler, T.M., The Use of Portable Instruments In Hazardous Waste Site Characterizations, Proceedings of the National Conference on Management of Uncontrolled Hazardous Waste Sites, Hazardous Materials Control Institute, Silver Spring, Maryland, 1982.

Cox, R.D., Sample Collection and Analytical Techniques for Volatile Organics in Air, Measurement and Monitoring of Noncriteria (Toxic) Contaminants in Air, ed. E.R. Frederick, Air Pollution Control Association, Pittsburgh, 1983.

Daggett, D.L., and Stock, T.H., An Investigation into the Storage Stability of Environmental Formaldehyde Samples, Am. Ind. Hyg. Assoc. J., vol. 46, no. 9, pp. 497-504, 1985.

Fletcher, R.A., A Review of Personal, Portable Monitors, and Samplers for Airborne Particles, J. Air Pollut. Control Assoc., vol. 34, pp. 1014-1016, 1984.

Fox, D.L., Air Pollution, Anal. Chem., vol. 57, pp. 223R-238R, 1985.

George, A.C., Passive, Integrated Measurement of Indoor Radon Using Activated Carbon, Health Phys., vol. 40, pp. 867-872, 1984.

George, A.C., and Breslin, A.J., Measurement of Environmental Radon with Integrating Instruments, Workshop on Methods for Measuring Radiation in and Around Uranium Hills, vol. 3, 9th ed., ed. E.D. Howard, Atomic Industrial Forum, Inc., Washington, D.C., 1977.

Girman, J.R., Hodgson, A.T., Robinson, B.K., and Traynor, G.W., Laboratory Studies of the Temperature Dependence of the Palmes Tube NO_2 Passive Sampler, LBL Report No. 16302 (ECB Vent 83-12), Lawrence Berkeley Laboratory, University of California, Berkeley, 1983.

Godish, T., Formaldehyde and Building-Related Illness, J. Environ. Health, vol. 44, no. 3, pp. 116-121, 1981.

Godish, T., Residential Formaldehyde Sampling--Current and Recommended Practices, Am. Ind. Hyg. Assoc. J., vol. 46, no. 3, pp. 105-110, 1985.

Katz, M., ed., Methods of Air Sampling and Analysis, 2d ed., American Public Health Association, Washington, D.C., 1977.

Katz, M., Advances in the Analysis of Air Contaminants: A Critical Review, J. Air Pollut. Control Assoc., vol. 30, no. 5, pp. 528-557, 1980.

Krost, K.J., Pellizari, E.D., Walburn, S.G., and Hubbard, S.A., Collection and Analysis of Hazardous Organic Emissions, Anal. Chem., vol. 54, pp. 810-817, 1982.

Lamb, S.I., Petrowski, C., Kaplan, I.R., and Simoneit, B.R.T., Organic Compounds in Urban Atmospheres: A Review of Distribution, Collection and Analysis, J. Air Pollut. Control Assoc., vol. 30, no. 10, pp. 1098-1115, 1980.

Lewis, R.G., and MacLeod, K.E., Portable Sampler for Pesticides and Semivolatile Industrial Organic Chemicals in Air, Anal. Chem., vol. 54, pp. 310-315, 1982.

Matthews, T.G., and Howell, T.C., Visual Colorimetric Formaldehyde Screening Analysis for Indoor Air, J. Air Pollut. Control Assoc., vol. 31, no. 11, pp. 1181-1184, 1981.

Matthews, T.G., and Howell, T.C., Solid Sorbent for Formaldehyde Monitoring, Anal. Chem., vol. 54, pp. 1495-1498, 1982.

Miksch, R.R., Anthon, D.W., Fanning, L.Z., Hollowell, C.D., Revzan, K., and Glanville, J., Modified Pararosaniline Method for the Determination of Formaldehyde in Air, Anal. Chem., vol. 53, pp. 2118-2123, 1981.

Nagda, N.L., Fortmann, R.C., Koontz, M.D., and Rector, H.E., Comparison of Instrumentation for Microenvironmental Monitoring of Respirable Particulates, Proceedings of the 78th Annual Meeting of the Air Pollution Control Association, Paper No. 85-30A.1, Pittsburgh, 1985.

National Institute for Occupational Safety and Health (NIOSH), NIOSH Manual of Analytical Methods, 2d ed., U.S. Department of Health, Education, and Welfare, Cincinnati, 1977.

Palmes, E.D., Gunnison, A.F., DiMattio, J., and Tomczyk, C., Personal Sampler for Nitrogen Dioxide, Am. Ind. Hyg. Assoc. J., vol. 37, pp. 570-577, 1976.

Riggin, R.M., Technical Assistance Document for Sampling and Analysis of Toxic Organic Compounds in Ambient Air, Report No. EPA-600/4-83-027, U.S. Environmental Protection Agency, Research Triangle Park, North Carolina, 1983.

Sawyer, R.N., and Spooner, C.M., Sprayed Asbestos-Containing Materials in Buildings: A Guidance Document, Report No. EPA-450/2-78-014 (OAQPS No. 1.2-094), U.S. Environmental Protection Agency, Research Triangle Park, North Carolina, 1978.

Solomon, W.R., Sampling Techniques for Airborne Fungi, Mould Allergy, eds. Y. Al-Doory and J.F. Domson, Lea and Ferbiger, Chapter 4, pp. 41-65, Philadelphia, 1984.

Spittler, T.M., Siscanau, R.J., Lataile, M.M., and Parles, P.A., Correlation Between Field GC Measurement of Volatile Organics and Laboratory Confirmation of Collected Field Samples Using the GC/MS, Proceedings of Specialty Conference on Measurement and Monitoring of Non-Criteria (Toxic) Contaminants, ed. E.R. Frederick, Air Pollution Control Association, Pittsburgh, 1983.

Turner, W.A., Spengler, J.D., Dockery, D.W., and Colome, S.D., Design and Performance of a Reliable Personal Monitoring System for Respirable Particulates, J. Air Pollut. Control Assoc., vol. 29, 7th ed., pp. 747-749, 1979a.

Turner, W.A., Spengler, J.D., Dockery, D.W., and Colome, S.D., Design and Performance of a Reliable Personal Monitoring System for Respirable Particulates, Proceedings of the Workshop on the Development and Usage of Personal Exposure Monitors for Exposure and Health Effects Studies, Chapel Hill, North Carolina, 1979b.

Turner, W.A., Marple, V.A., and Spengler, J.D., Indoor Aerosol Impactor, Presented at the First International Aerosol Conference, Minneapolis, 1984.

U.S. Environmental Protection Agency (EPA), Workshop on Indoor Air Quality Research Needs, Interagency Research Group on Indoor Air Quality, EPA, Washington, D.C., 1981.

Vo-Dinh, T., Gamaft, R.B., and Martinez, P.R., Analysis of a Workplace Air Particulate Sample by Synchronous Luminescence and Room-Temperature Phosphorescence, Anal. Chem., vol. 53, pp. 253-258, 1981.

Vo-Dinh, T., Development of a Dosimeter for Personnel Exposure to Vapors of Polyaromatic Pollutants, Environ. Sci. Technol., vol. 19, no. 10, pp. 997-1003, 1985.

Wallace, L.A., and Ott, W.R., Personal Monitors: A State-of-the-Art Survey, J. Air Pollut. Control Assoc., vol. 32, pp. 601-610, 1982.

Appendix C
UNITS AND SYMBOLS

This book conforms with conventions and rules espoused by the International System of Units (i.e., Systéme International--SI). SI has been adopted by most of the scientific community because it is coherent and simple. However, a large body of the technical literature has in the past used various mixtures of English and metric units.

This appendix clarifies the units used in this book and provides conversion factors for those readers delving into non-SI literature.

TABLE C-1. SI Base Units

Quantity	Name	Symbol
length	meter	m
mass	kilogram	kg
time	second*	s
electric current	ampere	A
temperature	kelvin	K
luminous intensity	candela	cd
amount of substance	mole	mol

*Also minute (min), hour (h), day (d), week (wk), and year (y).

TABLE C-2. Examples of Derived Units

Quantity	Name	Symbol	In terms of SI base units
area	square meter		m^2
density (mass)	kilogram per cubic meter	ρ	kg/m^3
energy	joule	J	$m^2 \cdot kg/s^2$
force	newton	N	$kg \cdot (m/s^2)$
frequency	hertz	Hz	$1/s$
pressure	pascal	Pa	$(1/m) \, kg \, (1/s)$
velocity	meter per second		m/s
volume*	cubic meter		m^3
activity of a radionuclide	becquerel	Bq	$1/s$
air exchange rate	air changes per hour	ν	$(m^3/h)1/m^3$
volume flow	cubic meter per second		m^3/s

*Also liter (L).

TABLE C-3. SI Prefixes

Factor	Prefix	Symbol	Factor	Prefix	Symbol
10^{18}	exa	E	10^{-3}	milli	m
10^{15}	peta	P	10^{-6}	micro	μ
10^{12}	tera	T	10^{-9}	nano	n
10^{9}	giga	G	10^{-12}	pico	p
10^{6}	mega	M	10^{-15}	femto	f
10^{3}	kilo	k	10^{-18}	atto	a

TABLE C-4. Some Useful Conversion Factors

Multiply	By	To obtain
ft	0.3048	m
ft^3	28.3	L
ft^3/min or cfm	0.472	L/s
in of water (60 °F)	249	Pa
psi	6.89	kPa
mi/h (mph)	0.447	m/s
calorie (g)	4.19	J
ft/min or fpm	0.00508	m/s
in of mercury (60 °F)	3.38	kPa
km/h	0.278	m/s
mi	1.61	km
WL (working level)	2.1×10^{-5}	J/m^3
Ci	3.7×10^{10}	Bq
L	10^{-3}	m^3
ft^2	0.0929	m^2
electronvolt (eV)	6.25×10^{18}	J
ppm (by volume at 25 °C, 101 kPa)	(molecular weight)/24.45	mg/m^3

INDEX